About Demos

Who we are

Demos is the think tank for everyday democracy. We believe everyone should be able to make personal choices in their daily lives that contribute to the common good. Our aim is to put this democratic idea into practice by working with organisations in ways that make them more effective and legitimate.

What we work on

We focus on six areas: public services; science and technology; cities and public space; people and communities; arts and culture; and global security.

Who we work with

Our partners include policy-makers, companies, public service providers and social entrepreneurs. Demos is not linked to any party but we work with politicians across political divides. Our international network – which extends across Eastern Europe, Scandinavia, Australia, Brazil, India and China – provides a global perspective and enables us to work across borders.

How we work

Demos knows the importance of learning from experience. We test and improve our ideas in practice by working with people who can make change happen. Our collaborative approach means that our partners share in the creation and ownership of new ideas.

What we offer

We analyse social and political change, which we connect to innovation and learning in organisations. We help our partners show thought leadership and respond to emerging policy challenges.

How we communicate

As an independent voice, we can create debates that lead to real change. We use the media, public events, workshops and publications to communicate our ideas. All our books can be downloaded free from the Demos website.

www.demos.co.uk

First published in 2006
© Demos
Some rights reserved – see copyright licence for details

ISBN 1 84180 162 3
Copy-edited by Julie Pickard, London
Typeset by utimestwo, Collingtree, Northants
Printed by Upstream, London

For further information and
subscription details please contact:

Demos
Magdalen House
136 Tooley Street
London SE1 2TU

telephone: 0845 458 5949
email: hello@demos.co.uk
web: www.demos.co.uk

The Slow Race
Making technology work for the poor

Melissa Leach
Ian Scoones

DEM⊙S

DEM✪S

Contents

Acknowledgements

This pamphlet has built on insights from work with colleagues and partners in developed and developing countries alike over several years. For their contributions to the ideas here, we would especially like to thank members of the Knowledge, Technology and Society (KNOTS) team at the Institute of Development Studies (IDS), colleagues at SPRU (Science and Technology Policy Research) and members of STEPS (Social Technological and Environmental Pathways to Sustainability), our new joint IDS-SPRU research centre at the University of Sussex, funded by the Economic and Social Research Council (ESRC). Our hope is that this centre will build a new network to take forward intellectually and practically some of the challenges outlined here.

For further insights and cases, and for support to our thinking with partners around issues of citizen engagement with science, our thanks are due to members of the Science and Citizens Programme of the Department for International Development-funded Development Research Centre on Citizenship, Participation and Accountability, based at IDS, and to the ESRC Science in Society research programme which helped support its meetings. Further colleagues, funders and partners in projects that helped inspire the pamphlet's arguments and cases are too numerous to mention, but we are particularly grateful to Gordon Conway, Gary Kass and Andrew Scott for their comments on an early draft. Thanks also to Jack Stilgoe, James Wilsdon and the staff of Demos for seeing the pamphlet through to publication. Finally, though, the pamphlet would not have happened without the support of the Rockefeller Foundation for necessary discussions and writing time through a project on 'governing technologies'.

Melissa Leach
Ian Scoones
June 2006

About the authors

Melissa Leach is a social anthropologist and professorial fellow at the Institute of Development Studies, University of Sussex, where she leads the team on Knowledge, Technology and Society (KNOTS). Her research and publications have focused on science–society relations, knowledge and policy processes around environmental and forest issues, natural resource management, health technologies and vaccines, both in West Africa and the Caribbean, and comparatively in the UK.

Ian Scoones is a professorial fellow at the Institute of Development Studies, where he is a member of the KNOTS team and convenes its agriculture programme. With a background in natural resource ecology, he has worked extensively at the interface between the natural and social sciences, exploring the relationship between science, policy, and local knowledge and practice in relation to agricultural biotechnology, soils management, pastoralism and dryland agriculture, largely in sub-Saharan Africa.

1. Three races

At no point in our whole history has the speed and scale of technological change been so fast and pervasive.

Gordon Brown[1]

The science races are on. After decades of relative neglect, science and technology are firmly back on the international development agenda. Science is woven into the UN's eight Millennium Development Goals.[2] The 2005 report from the Commission for Africa recommends that $3 billion should be invested in developing centres of excellence in science and technology.[3] New scientific initiatives such as the GFATM (Global Fund to Fight AIDS, Tuberculosis and Malaria) are emerging and attracting funding through schemes such as the International Finance Facility championed by Gordon Brown.

In Dakar in September 2005, Yaye Kene Gamassa Dia, the president of the recently created Committee of African Science Ministers, spoke of the need for 'a new vitality in the scientific and technological systems of African countries'.[4] In the UK, the Department for International Development (DFID) appointed a chief scientific adviser to produce a science strategy for development.[5]

All of this interest comes on the back of rapid advances in IT, biotechnology and nanoscience. These hold out the promise of new drugs, vaccines and seeds, and generate claims of breakthroughs that could solve poverty, illness and environmental decline. But what does this re-emergence onto the international stage of science and technology really mean? Where will these global science races take us? Who will win the prizes, and who will be left behind? And will this extra investment in science and technology really work for the poor?[6]

Behind the breathless enthusiasm for linking science, technology and development lie some contrasting views of what this might involve. Policy debates are dominated by two global science races, each of which is said to be pushing the international community down a particular path. This pamphlet explores the pros and cons of these two races and suggests that a third, less glamorous, but ultimately more important race is being overlooked.

The race to the top in the global economy

For many, science and technology are seen as a spur to economic growth in an increasingly competitive global economy. This fits with a view of development as modernisation, presuming that developing countries will move through a series of stages towards industrial and postindustrial glory. Science and technology help speed particular countries through these stages, and can perhaps enable them to 'leapfrog'. It is assumed that poverty will be reduced by the trickle down of economic benefits to the poor. The 'Asian tigers' and the exploding economies of China and India provide the models. Here, growth and poverty reduction seem to go hand in hand.

The UN Millennium Project report on science, technology and innovation argues that 'creating incentives and promoting an enabling environment for foreign direct investment is one of the most important mechanisms for building technological capacity'.[7] As Harvard professor Calestous Juma put it in a recent essay: 'A new economic vision . . . should focus on the role of knowledge as a basis for economic transformation.' Listing the requirements for Africa to 'go for growth', he identifies renewing infrastructure, building human capabilities, stimulating business development and increasing participation in the global economy.[8]

The race to the universal fix

For others, the race is for breakthroughs in science and technology that will have a direct and widespread impact on poverty. The prizes here are big-hitting technologies with the potential for global scope and applicability. Development models include the so-called green revolution, the spread of high-yielding varieties of crop staples, particularly across Asia in the 1960s and 1970s, and the mass vaccination campaigns that eradicated smallpox in the 1960s.

This race fits with a view of development as a matter of common interest and global responsibility, where science and technology are directed towards the problem of poverty. Moreover, in an increasingly interdependent world where people and microbes move freely, neglecting such development creates threats – whether in terms of the spread of disease, transborder environmental damage, or growing insecurities. It is thus in all our interests, it is argued, to invest in science and technology to avert these dangers.

In the past, the iconic model for this was investment in public science. Today, though, the talk is of development partnerships between the public and private sectors. This is exemplified by the efforts of Bill and Melinda Gates, whose Grand Challenges in Global Health are 'a major effort to achieve scientific breakthroughs against diseases that kill millions of people each year in the world's poorest countries'. The ultimate goal of more than US$430 million of funding is to create technologies 'that are not only effective, but also inexpensive to produce, easy to distribute, and simple to use in developing countries'. As Dr Elias Zerhouni, director of the National Institutes of Health, confidently predicts: 'Many of these research projects will succeed, leading to breakthroughs with the potential to transform health in the world's poorest countries.'[9]

The slow race to citizens' solutions

These first two races may grab the headlines. But there is a third, alternative race, which also demands our attention. This emphasises pathways to poverty reduction which may involve science and technology, but are specific to local contexts. It recognises that technological fixes are not enough, and that social, cultural and institutional dimensions are also key. And it sees science and technology as part of a bottom-up, participatory process of development, where citizens themselves take centre stage. Rather than being viewed as passive beneficiaries of trickle-down development or technology transfer, in this race, citizens are seen as knowledgeable, active and centrally involved in both the 'upstream' choice and design of technologies, and their 'downstream' delivery and regulation.

The models here include the appropriate technology movement and approaches to participatory technology development which became popular in the 1970s. Debates about public engagement with science have intensified in Europe over the past few years, as highlighted in recent Demos pamphlets.[10] Yet discussion of how these debates translate into the global development arena is only now emerging.[11]

These three races are not mutually exclusive and all are important. Most aid agencies highlight all three at different times and for different purposes.[12] But there are trade-offs and tensions between them, with serious implications for investment and governance. Making science and technology work for the poor is no straightforward task.

This is not the first time that the relationships between science, technology and development have been discussed. Debate on these topics within the United Nations system began over 40 years ago in 1963, when the first United Nations Conference on the Application of Science and Technology for the Benefit of the Less Developed Countries met in Geneva. In 1972, the landmark UN Conference on the Human Environment in Stockholm underlined the potential role of technology in both creating and solving the interlinked problems of development and environment. It recommended large-scale support to science and technology that could help solve developing countries' problems. Seven years later the UN Conference on Science and Technology for Development in Vienna attempted, inconclusively, to reach agreement on global principles and institutional arrangements to guide and facilitate this.[13] It established an

Intergovernmental Committee on Science and Technology for Development, which, in the late 1980s after disappointing progress, was transformed into a Commission on Science and Technology for Development within the UN's Economic and Social Council (ECOSOC).

These international deliberations had a strongly global, top-down flavour, emphasising the transfer of technologies, resources and advice to poorer countries. But at the same time another set of discussions produced alternative, bottom-up visions of development and technology. In the 1970s, EF Schumacher's landmark book *Small is Beautiful* sparked debate and action with its notion of 'appropriate technology'.[14] Schumacher himself founded the Intermediate Technology Development Group (ITDG), now Practical Action, which works to enable poor people to develop skills and technologies which give them more control over their lives and contribute to the sustainable development of their communities. Visions of the role of science and technology in development have always been diverse, as diverse as visions of development itself.

An outline of the pamphlet

In this pamphlet, we ask the following questions: What makes science and technology work for the poor? What roles might technologies play in the futures of people in developing countries? What forces will be involved in shaping this? How can poor people become more involved in shaping their own technological futures? And how can those who work in science and development – as researchers, aid donors, policy-makers – help? In particular, how should science aid spending best be directed?

Our central argument is that the slow race to citizens' solutions needs to be given more attention. Our argument unfolds through the next six chapters. Chapter 2 takes a hard look at the 'race to the top' and the 'race for the universal fix'. We argue that the neglect of poor people's own priorities in a trickle-down model of development may be that model's Achilles' heel. Similarly, the tendency towards universalised views of poverty problems means that technical fixes can miss their mark badly if we ignore poor people's own perspectives and concerns. Simple storylines may be useful to mobilise global resources, but their spending needs to link global initiatives to local definitions of problems and solutions.

Chapter 3 argues that approaches to innovation need to be rooted in these local realities. Linear, technology-transfer approaches often fail. Instead, a more participatory approach is needed, where innovations are seen as part of broader systems of governance and markets, extending from local to national and international levels.

Chapter 4 focuses on these issues of governance, looking in particular at technology access and ownership. The increasing dominance of the private sector may lead science and technology away from pro-poor, locally embedded priorities. Emerging arrangements such as public–private partnerships offer much promise in redressing this balance, but they also need to be evaluated critically, and in many instances are no substitute for publicly funded initiatives.

Chapter 5 turns to how risks and uncertainties arising from technological applications should be regulated. Developing country settings often pose huge challenges for regulation, which cannot proceed through the transfer of models from Europe and the USA. Instead, innovative approaches to 'inclusive regulation' are needed. At the same time, international regulatory mechanisms need to rethink their assumptions about 'sound science' and allow developing country agendas onto the negotiating table.

The core challenge is how to involve people, especially marginalised people, in decisions about innovation and technology. Chapter 6 argues that this requires a new vision of citizenship that goes beyond 'public engagement with science'. Rather, it needs more active engagement with broader questions about how science and technology agendas are framed, the social purposes they serve, and who stands to gain or lose from these. We offer several examples of science and technology in developing country contexts, some of which are signposts, and others which are diversions in the global science races.

In the final chapter, we ask what the next steps should be in the slow race to citizens' solutions. We look at ways of enabling citizen involvement in defining research priorities, in responding to regulatory challenges, in creating a new generation of science and technology development professionals and in organising the spending of development aid. To this end, we recommend the piloting of a series of 'citizens' commissions for science and technology futures'.

This pamphlet comes at a time when the promise of science for development is attracting wide-eyed policy attention. But as we have seen excitement grow, we have seen the views and settings of real people in developing countries overlooked. By drawing on academic research – our own and others' – and experience in developing countries, we want to remind policy-makers, scientists, social scientists and non-governmental organisations (NGOs) of the missing voices in this debate. We are driven by a wish to see institutions make the most of recent interest and investment in science, technology and development. Our argument is that citizen engagement is vital if science and technology are to respond to the challenges of international development. Our hope is that this argument for citizens' solutions provides a contribution to enrich and run alongside the existing debate.

2. Whose problems? Whose solutions?

The two science races that capture most international attention both suggest pathways to poverty reduction. But each also raises questions about the ways problems and solutions are being defined, suggesting limitations and trade-offs in making technology work for the poor.

The race to the top in the global economy holds that – as the Millennium Project's Task Force on Science, Technology and Innovation put it – 'creating links between knowledge generation and enterprise development is . . . one of the greatest challenges facing developing countries'.[15] It sees technology and innovation in promoting industrial production as the basis for development, enabling countries to move away from dependence on natural resource extraction. In the African context, the Commission for Africa recommended the establishment of a network of centres of excellence within Africa to help the continent catch up and keep up with the pace of technology-led economic growth. But this hopeful vision raises some tricky questions. Who is all this innovation and technology for? Do the benefits necessarily trickle down to the poor? Is hi-tech industrial-led development going to generate jobs for the relatively unskilled?

Take the example of Bangalore in south India, a city that deeply impressed Tony Blair when he visited in 2002. A few years later he recalled:

I remember sitting in a brand-new state-of-the-art university complex in Bangalore, talking to leading biotech entrepreneurs, many of them women academics who had branched out into business, confidently predicting they would beat Europe hands down in the biotech business in a few years.[16]

The growth of the IT sector in Bangalore, and India more broadly, has become legendary. The software boom, followed by an array of spin-offs, has fuelled year-on-year growth in the national economy, with more and more investors seeing India as a desirable destination for investment. Bangalore itself is seen as a 'knowledge industry' hub, combining a highly educated yet low-cost workforce with top-class research establishments. Tax breaks from government and routes to avoid overwhelming bureaucratic hurdles have increased the attraction still further.[17] India's National Association of Software and Service Companies (NASSCOM) estimates that a million jobs have been created to date, a figure which is expected to double over five years, especially if IT-enabled call centre jobs are added in.

If we take a wider developmental view, however, there are downsides to the Bangalore success story. Technology industries may bring riches and jobs for some, but not everyone. A million new jobs is good news, but remains a drop in the ocean compared with the 470 million-strong labour pool in India. Urban infrastructure is under extreme pressure as rapid growth puts pressure on transport, water, waste and other services. Enclaves for the technology elite have been built, while the rest of the city suffers from a lack of investment in basic infrastructure.

And the rural hinterlands, where poverty is still pervasive, receive little benefit from the urban, technology-led boom. A recent spate of farmer suicides in the areas around Bangalore has brought these two worlds into sharp relief, and highlighted how the crisis in agricultural and livelihood systems is not being addressed by the technology boom. In a speech to the Indian Science Congress in January 2006, Prime Minister Manohman Singh and Nobel laureate Amartya Sen both pointed to the need for balanced social and economic development and a science and technology thrust that meets the needs of the rural poor.[18] In Bangalore the technology elite profit from the knowledge economy. But at the same time the broader populace – over 60 million people in Karnataka state – see little of the economic boom that Tony Blair so fears. Reaping the gains of globalisation is possible for some, not others. And the predicted trickle down is hard to find.

Technology-led economic growth can play an important role in development. But on its own, it will not be enough to meet the Millennium Development Goals. A failure to pay attention to inclusive development may produce repercussions that undermine economic progress in the long run. Reducing poverty requires complementary approaches.

Can we fix it?

The race to the universal fix portrays technological breakthroughs as a direct route to combating poverty. Between 1995 and 2004, the UK government invested £190 million in support of poverty-focused technology development in renewable natural resource management and agriculture, and has plans to invest more in a new Strategy for Research on Sustainable Agriculture in developing countries that will run until 2016.[19] British research councils have directed a significant proportion of their spending towards science and technology aimed at poverty-related applications.

For example, the Medical Research Council (MRC) is currently funding major programmes of work on poverty-related disease including HIV/AIDS, tuberculosis and malaria, and childhood infections in Africa, as well as research programmes in China, India and Jamaica. The MRC has a concordat with DFID to coordinate policies and share resources for research into the health of developing societies, which amounted to £22.5 million per annum in 2002/03. At a global level, the 15 international centres within the Consultative Group on International Agricultural Research (CGIAR) receive £220 million every year, and DFID has just doubled its contribution to these to £20 million per annum as part of its research funding framework.[20]

Similarly, the GFATM has so far raised $2.2 billion in pledges from a mix of governmental and private philanthropic donors, and has disbursed $1.5 billion of this to 160 programmes in 85 countries in its first and second funding rounds. Much of this investment is justified by the prospect of 'big hit' technologies with potential for massive impact. Software billionaire-turned-philanthropist Bill Gates has begun targeting the money in his foundation with 14 'grand challenges' for research in global health (see box 1).

Box 1. Grand challenges in global health[21]

Initiated and managed by the Bill and Melinda Gates Foundation, together with the Wellcome Trust, Foundation for the National Institutes of Health and Canadian Institutes of Health Research, a scientific board identified 14 challenges to serve seven long-term goals to improve health in the developing world:

Improve childhood vaccines
1   Create effective single-dose vaccines.
2   Prepare vaccines that do not require refrigeration.
3   Develop needle-free vaccine delivery systems.

Create new vaccines
4   Devise testing systems for new vaccines.
5   Design antigens for protective immunity.
6   Learn about immunological responses.

Control insects that transmit agents of disease
7   Develop a genetic strategy to control insects.
8   Develop a chemical strategy to control insects.

Improve nutrition to promote health
9   Create a nutrient-rich staple plant species.

Improve drug treatment of infectious diseases
10   Find drugs and delivery systems to limit drug resistance.

Cure latent and chronic infection
11   Create therapies that can cure latent infection.
12   Create immunological methods to cure latent infection.

Measure health status accurately and economically in developing countries
13   Develop technologies to assess population health.
14   Develop versatile diagnostic tools.

Funding for poverty-related science and technology, impressive as it may seem, is currently dwarfed by the resources devoted to technologies for commercial markets – a problem that we discuss in the next chapter. Nevertheless, even current levels of investment in 'big hit' technologies, together with their high profile in the media, present a beguiling idea that there are technological fixes for the problems of the developing world. Scientists have become well versed in making their case to big funders, often promising more than is realistically deliverable. Where such initiatives claim to be attuned to poverty reduction, problems and solutions are often framed in universalised terms – applicable anywhere, at any time. The nature of the health, food or agricultural problem is assumed to be broadly similar across vast areas, so that technological solutions can be transferred unproblematically and applied at scale.

Such universalised talk of 'big-hitting' technological solutions runs into several problems. First, ecologies and the practices that people have developed to sustain their livelihoods are highly diverse. The particular interactions between social and ecological change vary across regions, localities and sometimes even within communities, producing multiple patterns and multiple needs. A one-size-fits-all solution is often inappropriate, and a magic bullet will often miss its mark. Second, hype about new technologies may obscure important opportunities to spread already tried and tested 'old' technologies, adapting these to particular local circumstances. So, for example, hybrid seeds, perhaps enhanced by biotechnological techniques of marker-assisted selection, may have more impact than untried genetically modified (GM) varieties. Third, problems of poverty, hunger and illness are not just the result of technical matters. Just as important may be failures of markets, unequal social relations, political questions, conflict and other issues. If one looks at local livelihood systems, it becomes clear that the social, political and technical are intimately intertwined. In this context treating science and technology as a separate issue is dangerous. The history of technology development is full of examples of technologies that look good on paper, but never get beyond laboratory benches or abandoned field trials.

Box 2. Problems of the universal fix?

Addressing soil fertility challenges[22]

One of the suggested 'quick wins' identified by the UN Millennium Project was a massive effort to replenish nutrient-depleted soils in Africa through a combination of chemical fertilisers and agro-forestry. Nigeria is hosting a 'Fertiliser Summit' for Africa in 2006, with President Obasanjo taking a lead on these recommendations. With less than 10 kg per hectare (ha) of fertiliser being applied on average to African farm lands, access to soil nutrients is a key issue if African agriculture is to grow sustainably. But what should be done? Pedro Sanchez and Jeffrey Sachs of the Millennium Project argue for a major technology-led effort. But Africa has seen many fertiliser and soil management projects falter in the past. Grand plans for 'soil replenishment' miss the need for a balanced approach, which recognises the diversity of soil across farm landscapes. Instead, policies should be attuned to local soils, markets and farming conditions, and take a decentralised, partici-patory approach. It is not just a lack of nitrogen or phosphorous that is the issue. There is a wider set of market, institutional and policy matters that need to be addressed.

Responding to scarcities[23]

Many technological solutions are justified in terms of scarcity. Better, bigger, more efficient technologies can, it is argued, be engineered to respond before things run out. The contemporary 'crises' of energy, water and even food are often constructed in these ways. Debates about water resources in the developing world have long been thought about in these terms. Grand schemes are designed on the basis of hydrological projections of need – for urban consumption or irrigation development. Big dams, river diversions, catchment transformation, piping and storage systems are the typical engineering solutions.

In the dry zones of Kutch in Gujarat, India, farmers approach the issue of water scarcity in a different way. First, there are multiple scarcities – it depends on the place, the time and the purpose to which the water is being used. There is huge uncertainty and a number of ways of responding to the situation. There is not one solution, but many. And the issue is not so much about absolute amounts of water, but about its distribution. Who gets access, and when? The discussion covers wider issues of access, rights and the politics of water. It is not that engineering science cannot contribute, but the framing of the questions, the understanding of the problems, and the way technologies are designed as solutions may be very different.

Building demand for vaccines[24]

Childhood vaccines are seen as vital solutions to diseases of poverty. Policy-makers and health professionals assume that building demand for vaccine technologies requires knowing about the specific diseases that vaccines prevent, and this is often the focus of health education programmes. Yet whether or not parents and carers demand vaccines for their children can depend more on how vaccination fits in the contexts of people's lives, including perceptions around child health and illness.

In the Gambia, for example, most mothers see the value of infant vaccines but they don't distinguish vaccinable from non-vaccinable diseases, or protection from cure, in the same terms as biomedicine. Instead, they understand vaccines to promote infant health generally by building up strength and weight, staving off illness, and 'chasing out' illness from a child's body. As one mother put it: 'The injection strengthens the health of the child. It gives the child a good body.' Many mothers feel that vaccinations are effective against illness in general, especially the 'small illnesses' that afflict children. In a survey, 29 per cent of urban and 48 per cent of rural mothers could name no biomedically 'correct' vaccinable diseases, yet were actively seeking vaccination, which most saw as a complement to, rather than replacement for, 'traditional' practices, such as the Islamic 'talismans' that many tie on their children's bodies for protection. Understanding why people accept (and sometimes reject) technologies such as vaccines requires engaging with local cultural meanings that may differ strongly from those of scientists and policy-makers.

This is not to say that the technologies highlighted in box 2 are not potential development solutions. Fertilisers, irrigation technologies and vaccines clearly have key roles to play. However, they may miss their target for reasons that are poorly understood by those promoting them. The implication is that technological choices and strategies for promoting technology uptake have to be attuned to local livelihoods, knowledge and social impacts. This in turn has implications for the ways in which technologies are appraised and priorities for investment chosen. Making the case for a slower approach involving poor people themselves, Miriam Were, Board Chair of the African Medical and Research Foundation (AMREF), remarks that:

Most donors are in a rush. They are in a hurry to have results tomorrow, and they are in a hurry to define the specificity of projects. As a result, projects often ignore important contextual issues. Donors, and we Africans ourselves, must recognise the local context and realize that this context may not be in support of the long-term benefits anticipated from the project we have designed our way. For instance, people say things like if we can produce this result within three years we can reduce poverty. Many of our people live in absolute poverty so they will not say no, even if what we suggest doesn't make sense to them. They will in fact get involved knowing very well that this thing won't last. We must address the issue of context much more carefully, and of course the overriding issue is poverty. Many of our people are living in absolute poverty of less than one dollar per day, and people who don't live with that don't quite understand what that means.[25]

A key challenge in the slow race to citizens' solutions is to enable local perspectives and experiences – those of people whose lives consist of 'getting by' in absolute poverty – to help shape spending priorities. As Miriam Were emphasises, this takes time and patience. It also requires innovation in priority-setting, and the development of new technological and social appraisal methodologies. For these to work, organisations funding or delivering technologies need to rethink how they bring local realities to bear on their efforts. This may mean some fundamental reconfiguring of aid agencies and research organisations, in ways that we outline in the final chapter.

3. Rethinking innovation

The 'universal fix' view holds that the technologies developing countries need already exist in other places, so the key task is to ensure their effective transfer. This approach has dominated research and development (R&D) systems for decades. Yet impacts have been uneven and insufficient to address the pressing needs of poorer people, particularly in marginal environments. The green revolution of the 1960s and 1970s brought huge gains in agricultural productivity for some, but its packages of high-yielding seeds, fertilisers and other inputs proved inappropriate for the ecological and socio-economic circumstances facing many small-scale farmers. Africa, in particular, missed out on the green revolution gains, in large part because of poor infrastructure, lack of irrigation and the complexity of the farming systems that had developed to deal with complex, diverse and risk-prone environments. Some now argue for a new technological revolution for African farming centred on bio-technologies, including GM crops, but such a strategy is likely to run into similar problems.

It was in response to many of the failures of technology transfer in the 1970s that new types of participatory technology assessment were initiated. These put farmers at the centre of the innovation process, working in collaboration with scientists to design new technologies and to adapt existing ones to local circumstances. These approaches recognised the value of local knowledge, moving away from the image of farmers as passive recipients of externally derived technology, to involve them as active, creative partners in technology development processes. So, for example, networks of farmers across Africa began to share ideas about soil and water conservation and soil fertility management techniques, linking with scientists in new ways and under different institutional arrangements. Several of the international agricultural research centres initiated programmes of participatory plant breeding, often involving social scientists to help broker the distinct conceptual and social worlds of farmers and scientists and to bring them together in new and productive ways.[26]

These participatory interactions have produced islands of success but have also raised questions about who controls the innovation process, and whose perspectives drive it. Too often, participation has meant simply co-option of local people into pre-set technological agendas. The huge imbalances in the power, reach and resources of local people and research agencies has contributed to this. Even where true collaborative arrangements have been established, these have often been dependent on the interest of key individuals and on temporary project funds, rather than being fully institutionalised in national and international innovation systems.

Banji Oyelaran-Oyeyinka, a Nigerian scholar based at the United Nations University in Maastricht, argues strongly that 'S and T policy design for development must be re-conceptualised in "systems terms" and take historical forces into account'.[27] This would represent a radical departure from the ways most external support has been organised to date. As John Mugabe, science and technology adviser to the New Partnership for Africa's Development (NEPAD), has put it:

> At the moment most of the funding to science and technology is project oriented . . . small amounts to short-term projects. . . . Many donors are still adopting a traditional project modality. Those who are serious about Africa's development need to move to institutional building.[28]

Instead of focusing on isolated examples, an innovation systems approach emphasises the networked interaction of multiple actors, public and private, local and national, in processes which initiate, import, modify and diffuse technologies.[29] It emphasises the links between these actors, which enable them to operate as an effective system, involving issues of funding, marketing, policy and legal frameworks. This contrasts strongly with a linear model in which funds are allocated to scientists, whose results are then handed to others, who are expected to deliver to a target audience.[30] DFID's crop post-harvest programme has been experimenting with innovation systems approaches in both Africa and Asia. Analysis from India shows that new patterns of partnership are slowly emerging to promote new 'pro-poor technologies' between different actors in the innovation system. In the area of post-harvest technologies, public and private research organisations are involved alongside NGOs, farmers' associations, processing and marketing businesses, linked to public sector research organisations and private sector firms. As John Mugabe argues, institutional relationships are critical to the success of a technological innovation, and its impact on poverty reduction.[31]

Many dimensions of innovation

Innovation should focus not only on the technology, but also on the social, cultural and institutional relationships that will enable the technology to work. There are numerous examples of technologies that already exist, which could have major impacts on poverty reduction, yet remain out of reach. To make existing technologies – sometimes very ordinary, everyday technologies – accessible to people living in poverty often means linking the technical with the social. For example, in parts of south Asia, a revolution in 'community-led total sanitation' has occurred as community organisation, empowerment and learning has facilitated the widespread building of low-tech, low-cost latrines.[32] To enable people to make use of technologies that may be available, but are poorly understood, often requires culturally appropriate communication strategies, improving people's knowledge and power to make technology choices. Far from resorting to the old-style 'deficit model', such approaches to science communication have the potential to network, empower and facilitate technology development in favour of the poor.[33]

An innovation systems framework that combines the technical with the social and institutional has been developed by ICRISAT (International Crops Research Institute for the Semi-Arid Tropics) and partners as a methodological tool for application in a wide array of technology development processes.[34] Embracing such an approach, especially combined with a real commitment to partnership and participation, brings many challenges to conventional professional hierarchies and institutional arrangements,[35] but these challenges will need to be faced by DFID and its partners over the coming years as its research strategy for agriculture, which uses an innovation systems approach, unfolds.

There is also a more fundamental challenge – to link local and national processes of innovation with global processes. There is much talk about building partnerships internationally for supporting science and technology development capacity in the south. But how will these initiatives fit with local and national innovation systems, particularly as many are premised on assumptions of linear technology transfer and infrastructure support? New thinking on these global partnerships needs to evolve, which takes imbalances in power relations between developing countries and international institutions seriously.

Proposals to create new scientific centres of excellence within developing countries – such as those recommended by the Africa Commission – may well be valuable ways forward. However, there will be a need to guard against an elitist form of science and technology development that looks upwards to its international funders and peers, rather than downwards to local contexts and users. The Kigali Institute for Science, Technology and Management (KIST) in Rwanda is one example of an African institution that strongly emphasises engagement with users.[36] Established in 1997 with donor support, the institute runs diploma and degree programmes which aim to build local and national scientific and technical capacity, while being linked to innovation projects focused on local needs. It has won awards for its small-scale technologies in the energy arena, such as improved bread ovens and biogas digesters.

Conceptualising the future of innovation systems is not just about building the 'hardware' of R&D infrastructure and capacity, but fundamentally about the 'software' of social and political relations among the many actors that are now involved, and the different interests that shape science and technology agendas. For example, with concentrations of innovation expertise increasingly held within multinational corporations, a major challenge is how to capitalise on this capacity and link it to wider systems which are attuned to the needs of the poor. This in turn raises questions about the governance of science and technology, which provide the focus of the next chapter.

4. Governing technologies

Past efforts to promote technology for development were based largely on a public sector model. The green revolution is a good example. This was supported through international agricultural research institutions linking with national research stations, with philanthropic support from US foundations. Many key advances in the health sector have been achieved through publicly funded laboratories, public research councils and philanthropic organisations such as the Wellcome Trust.

However, this is changing. According to the UN Conference of Trade and Development, private sector R&D in OECD (Organisation for Economic Co-operation and Development) countries made up over half of the $677 billion global R&D spend in 2002. A 2005 UNESCO report shows that Asian R&D investment, notably public R&D in China, is increasing, and now accounts for around 32 per cent of total global R&D spending. Asian countries are increasingly seen as an important destination for R&D by multinational corporations. In 1983 there were no such facilities in Asia, but by 2002 there were over 700. The R&D budgets of the top global life science companies now dwarf even those of international public research centres, let alone national research systems.[37] In the agricultural field, for instance, five large multinational companies – Bayer, Dow Agro, DuPont, Monsanto and Syngenta – spend $7.3 billion per year on agricultural research; more than 18 times the budget of the publicly funded CGIAR system.[38]

The squeeze on public budgets in developing countries has been intense, particularly following structural adjustment reforms. As a result there has been a brain drain of developing country scientists to international and private sector funded institutions. Even the ambitious plans to regenerate public sector R&D in Africa, such as the African Union's New Partnership for Africa's Development, which urges a commitment of $160 million to African innovation systems, will be a drop in the ocean.[39]

Unsurprisingly, private sector R&D is geared to markets where significant profits can be made. These are not in the poorer areas of the world, nor do they focus on the technologies that the poor need most. In the health field, a so-called '90:10' gap has emerged in which only 10 per cent of the overall world health research budget of $50–60 billion is spent on the diseases that affect 90 per cent of the world's population.[40] The drug market in Africa is such a tiny proportion of the global market that it can easily be dismissed by an 'accounting error'[41] to pharmaceutical companies with their sights fixed on the bottom line.

Public–private partnerships

Some view public–private partnerships (PPPs) as the answer. These aim to make private sector innovations available in areas where they would otherwise not be. A variety of mechanisms are being experimented with, including 'push' approaches where the public sector subsidises the costs of R&D in less profitable areas, and a variety of 'pull' approaches to create improved incentives for the development of PPPs. These include advance agreements by the public sector to purchase quantities of the product, tax breaks to corporations to encourage particular types of innovation, and patent extension deals. High-profile, publicly funded technology prizes have been suggested as a way of encouraging companies to focus at least some of their R&D effort on pro-poor technology development. Appealing to companies' sense of social responsibility, but also to the desire for future markets, other mechanisms focus on patent release arrangements, making otherwise patented products available at reduced cost in particular markets. Innovative thinking about further types of arrangement is needed. Box 3 offers some examples.

Box 3. Public–private partnerships

IAVI, the International AIDS Vaccine Initiative[42]

The initiative aims to further HIV/AIDS vaccine research worldwide, including the search for candidate molecules, the funding of clinical trials, work on delivery issues and wider policy and advocacy efforts. By 2004 IAVI had raised over $340 million. The aim is to work towards an effective and cheap vaccine which is available in poorer parts of the world. Vaccine development partnerships have been created between developing country organisations and northern research outfits, both public and private. The initiative spreads its funding across a diversity of players, and focuses on vaccine development and delivery issues rather than upstream research. It currently operates in 22 countries, and is increasingly decentralised in its operation, responding to early accusations of top-down, central control. The existence of regional offices and growing links with NGOs and civil society means the initiative is certainly broad-based. But despite its scale it still remains a small player in the overall HIV/AIDS technology innovation and delivery network, dwarfed by bigger funds, and much upstream research – both public and private – where such partnerships have little purchase.

East Coast Fever (ECF) vaccine project[43]

The ECF vaccine project is aimed at researching, designing and delivering a bioengineered vaccine against a parasite (*Theileria parva*) that has a significant impact on livestock in sub-Saharan Africa, particularly high-value exotic breeds and dairy animals. The project is based at the International Livestock Research Institute (ILRI), part of the CGIAR system based in Nairobi. DFID has provided

---

around £5 million of support. ILRI works together with the Kenya Agricultural Research Institute (KARI), the national agricultural research institute in Kenya and Merial, a French biotech company. Other university research groups are involved in particular research and monitoring aspects. The network developed so far has increased the capacity for innovation, clinical testing and delivery. But many questions remain, not least about its efficacy and demand for the vaccine if and when it is developed, particularly among poorer livestock producers keeping indigenous animals where ECF is not a priority. Whether a substantial investment of public donor funding was best invested in either this disease or this type of vaccine is certainly questioned by many in the field.

Striga-resistant maize[44]

The African Agricultural Technology Foundation (AATF) aims to broker arrangements to make innovative technologies available to poor farmers in Africa. Established in 2002, with funding from USAID, the Rockefeller Foundation and DFID, it works with African governments and a range of multinationals including Monsanto, Dupont, Dow Agro-Sciences and Syngenta. A flagship project, based in Kenya, has been working on striga-resistant maize, which has high demand among the smallholder areas of sub-Saharan Africa. Multiple actors are involved: the chemical company BASF has contributed patented genetics and a herbicide seed covering; AATF has brokered intellectual property rights sharing and assisted with regulatory approval; and CIMMYT (the International Maize and Wheat Improvement Center) and KARI (the Kenyan Agricultural Research Institute) have been involved in developing and adapting the product. The striga-resistant maize was released in mid-2005 to much fanfare and a number of Kenyan seed companies are now marketing the product.

A number of characteristics are evident from these examples. Each was initiated by seed funding from a development donor (in two of these cases, the Rockefeller Foundation); each involves a diverse group of actors; each relies on continued public support to keep going; each retains operating autonomy and has a governance structure in which both users and funders guide policy; and each has a strong and dynamic leadership and host organisation.

Experience with PPPs also reveals some problems. First, there is sometimes a naïve assumption that the private sector has off-the-shelf pro-poor technologies, or can rapidly switch its R&D systems towards these. Second, being focused on global and largely northern markets, private sector corporations are often dissociated from national and local innovation systems in developing countries. Thus the problems of a linear technology transfer model discussed in the last chapter arise when technologies are delivered. Third, where the focus is public subsidy of the upstream development of new technologies, then these may not find a market in developing countries. While advance purchase agreements may deal with this to some extent, questions arise about the long-term sustainability of such subsidies.

Seth Berkeley, CEO of the International AIDS Vaccine Initiative, commented in a recent *Newsweek* article: 'It's been difficult to convince the private sector to invest in vaccines – traditionally high-volume products that provide relatively low return on investment. Unlike a drug that patients may have to take for a lifetime, an effective vaccine is literally a 'one shot' deal – there are virtually no repeat customers.'[45] He went on to list a range of other factors that impeded engagement of the private sector in PPPs: 'Uncertain demand raises yet another obstacle. In the case of AIDS, poor villagers in Africa and India and their governments may not be willing or able to pay for an expensive vaccine, so the precise number of doses needed worldwide remains unclear.' He concluded that what is needed is more public funding – for research to 'overcome scientific roadblocks' – as enticements to 'bring the private sector to the table' in PPPs and, finally, liability protection and public insurance to offset risks for companies. This could add up to a large bill for public finances, one where trade-offs with other expenditures would have to come into play.

Similarly, Peter Jeffries, business development manager at Merial and a partner in the ECF vaccine PPP, highlighted a number of 'future challenges' for the initiative. His list includes intellectual property, continued development funding, regulatory clearance, manufacturing and supply and marketing. As he pointed out, these are standard issues for any vaccine, but the particular contexts where the ECF vaccine is supposed to be used are very different from the standard commercial setting. Indeed, the collapse of veterinary services in large parts of Africa and the challenges of delivery through decentralised, privatised systems is well known.[46]

Therefore, careful thought needs to go into which PPP-type arrangements might work for which technologies, products and settings. Global fund agreements may be appropriate for products such as vaccines, which are delivered on a large scale and require little local adaptation. But drugs or vaccines for particular regional disease variants, let alone crops for particular agro-ecologies, will require local adaptation, needing inputs from locally embedded research organisations, whether public or private. Technology prizes will work only where there is a major publicity gain to be had from developing technologies which are either high profile in themselves, or where there is competition between major companies. Patent release arrangements are problematic where technologies involve multiple patents on different component processes and products, owned by different companies. It took several years of negotiation and intervention by a number of influential players to get the vitamin A-enriched 'golden rice' available in the public domain because it was associated with more than 70 patents.[47] Such an effort is not going to be easily replicable for less attention-grabbing technologies.

Intellectual property rights

Some argue that the restrictions that intellectual property rights place on technologies are a real constraint to making them available to the poor, or to promoting public – as opposed to commercial – values. A recent survey of patents in *The Economist*[48] observed that over the past decade 'the number of patent applications has nearly doubled and continues to climb'. By conferring temporary rights (usually 20 years) the patent system in principle provides a time-limited commercial protection for innovators and their discoveries. In theory, the system encourages innovation and industrial growth, with specialisations emerging in upstream discovery and downstream commercialisation and delivery. As transferable rights in the market, patents are also key to company values, and the accumulation of patents can protect a firm from take-overs by pushing up perceived value.

But some commentators are critical of the patent system. James Boyle from Duke Law School argues that 'the current increase in intellectual property rights represents nothing less than a second "enclosure movement"'.[49] Exclusive rights, even if temporary, may restrict access to new technologies with prices being pushed up as a result. This may have significant impacts on the poor, and reduce the incentives to develop patentable technologies for such users. As technology solutions become ever more complex, the range of things that are 'novel, useful, non-obvious and man-made' increases, making managing – and enforcing – intellectual property arrangements increasingly difficult. As innovation takes place in more and more places, so the ability to manage intellectual property is undermined, despite the attempts of the World Trade Organization (WTO) and World Intellectual Property Organisation. The old world order where innovation occurred in the west and was licensed elsewhere is no longer the case. As the new and old tigers of east and south Asia gear up to compete in the global economy, new codes, rules and institutions will be required.

The past few decades have seen an expansion of intellectual property regimes globally, and an obligation for others to comply with free trade and intellectual property rules. But opposition to this system is building. The Adelphi Charter issued in London in late 2005 argues that a public interest test is needed before governments expand intellectual property rights. The UK Commission on Intellectual

Property Rights similarly argues that expanding patent regimes are not always good for the poor and for public-good innovation.[50] Others go further still. Building on the Linux software model, an increasingly influential group of researchers, business entrepreneurs and government officials, both north and south, are arguing for a 'patent commons', one that encourages 'open source' approaches,[51] where the restrictions of conventional intellectual property arrangements are replaced by flexible sharing of innovations.

In commercial systems characterised by growing complexity in innovation systems, processes of niche specialisation and convergence around particular technological solutions, the appeal of 'open innovation' systems is significant. These involve 'distributed peer production' across diverse individuals, organisations and places, linked to open licensing systems which allow different innovators to make adaptations of a core platform technology without restrictions. In the area of agricultural biotechnology, Australian scientist Richard Jefferson of Biological Innovation for Open Society (BiOS) argues that because few companies in the private sector hold the patents on crucial biotechnologies and processes, they are acting to 'dominate, then destroy the industry'. BiOS and others argue for parallel engineering where innovations are made available for both not-for-profit and for-profit research and product development.[52]

Technology entrepreneurs

The attention given to formalised public–private arrangements underestimates, at least in some areas, the potential for private entrepreneurs to go it alone. The extraordinarily rapid spread of mobile phone technologies throughout the world, including into remote and poverty-stricken parts of Africa, reminds us of the potential for private sector-led but demand-driven technology revolution. In some countries in Africa the mobile phone market is more than doubling every year. Here is evidence for what University of Michigan professor CK Pralahad has termed 'the fortune at the bottom of the pyramid', a market of over a billion people with annual incomes less than $1500.[53] For example, the impacts of the spread of mobile telephony are multiple. A London Business School/Vodafone study showed a strong correlation between expansion of mobile phone use and the growth of the economy over time.[54] There are also other benefits – in sending money to remote areas or in getting information on health, agriculture and livelihood issues. But for such impacts to be felt widely the conditions have to be right.

Those countries where mobile phone technologies have reached furthest have had a combination of a competitive market that has kept prices low and a policy environment that does not hike costs. But even at $20 or $30 per handset, the cost remains prohibitive for many, especially in poorer rural areas. A recent study in Uganda, for example, showed that despite skyrocketing mobile phone ownership, 90 per cent of the country, mostly in rural areas, do not have a connection. GrameenPhone in Bangladesh, working with a Norwegian telecoms operator, is developing a village-based approach for ensuring access where loans are offered and rental arrangements pay back the cost over time. Reaching poorer people has often depended on locally adaptive social arrangements: for instance to share phones or give poor neighbours access to those in richer households.[55] Parallels exist in the agricultural sector, where private companies produce a high-quality suite of hybrid seed products, which then spread rapidly, as they have for instance for maize in India. Where private companies can produce high-quality, appropriate products at a sufficiently low price for poorer people to afford, then arguably they should be left to get on with it.

Another route to getting good technologies to the poor has been through subversion of the mainstream, capitalist property rights regimes. What to some is technology theft or piracy, to others is the emergence of so-called 'Robin Hood' companies, copying patented products and selling them on to willing consumers at low cost. A Gujarat-based seed company made GM cotton available to farmers in India several years before Monsanto's product was formally approved for release.

Copies of drugs have been made available legally through well-developed reverse engineering techniques, mostly for off-patent technologies. Such strategies have been highly successful in producing cheap generic drugs for poor people. The Indian pharmaceutical industry has for many years been at the forefront of this. With 122,000 chemists and chemical engineers, who are relatively cheap to hire by global standards, the opportunities have been significant. Companies such as Ranbaxy and Dr Reddy's have become global players on the back of successful generic drug manufacture.[56] Generics manufacturers from India produce over half of the global supply of antiretroviral HIV/AIDS treatments, with low-cost supply focused on poorer countries. By the standards of the developing world these are now large, established companies: Ranbaxy's total global sales are now approaching $1.5 billion.

But how can we expect developing country science-based companies to compete in the cut-throat world of pharmaceuticals, or indeed any other area of science and technology?[57] Drug development is risky, with total costs for a new drug – from molecule discovery, to development, to clinical trials and regulatory approval – estimated at around $1 billion, and with the market focused almost exclusively on diseases and conditions of rich people in rich countries . Ranbaxy and Dr Reddy's, along with an array of other smaller Indian companies, find it hard to get a look in. With the approval of the Indian Patents Act in 2005, as part of a WTO agreement, there is a fear that Indian companies will abandon their niche in product re-engineering.[58] The likely upshot is that most Indian players will end up in outsourcing arrangements within a framework set by the big pharmaceutical companies. Such companies are increasingly prey to takeovers and mergers with large multinational corporations, who see them as useful sources of relatively cheap, outsourced skills and labour – the 'biotech coolies' of the international system. If this scenario unfolds, who will then provide the re-engineered generic drugs for wide, low-cost distribution in the developing world? Will the new global funding initiatives – such as the GFATM – step into the breach and provide the funding necessary? Or will a more unregulated, illegal parallel market emerge where quality, safety and efficacy are open to question? Will the result be a bewildering array of unregulated products that may impose wider risks?

Where the private sector cannot reach

Despite all this dynamism, it is clear that the private sector – even in alliance with innovative new forms of public support – cannot always make technology work for the poor. There will continue to be areas of technology and areas of the world where markets are thin or non-existent. Here there is no substitute for continued public investment on a substantial scale. The lesson from the green revolution is that only with major investments in infrastructure and research will significant gains be made. Public investments therefore need to be targeted strategically to those areas that will remain unattractive for the private sector. New mechanisms to enable this, such as Gordon Brown's International Finance Facility, are a step in the right direction.

5. Regulating risk

Science and technology bring promises and opportunities. But they also bring risks and uncertainties. Efforts to promote dynamic innovation systems need to run hand in hand with regulation. In developed and developing countries, public debate about BSE (bovine spongiform encephalopathy), GM crops and the MMR (measles, mumps, and rubella) vaccine has revealed a heightened sense of anxiety about the negative impacts of technologies linked to distrust in the governments and companies promoting them. Some have suggested that we are witnessing the emergence of a 'risk society', one where in late modernity the institutions of industrial society both produce and legitimate hazards that they cannot control.[59] There is a mismatch between the character of these hazards and people's culturally grounded reactions to them, and the scientific and bureaucratic apparatus charged with managing risk. In the process, society has become 'reflexive', compelled by this mismatch to question its foundational principles, potentially leading to public dissent and a questioning of the ways public institutions regulate.

But how applicable are notions of an emergent risk society in the developing world? This discussion can overstate the novelty of risk as a phenomenon, and the novelty of a mismatch between public perceptions and institutional regimes. Risks, hazards and un-certainties have long been experienced in developing country settings in the constant interplay of ecological and bodily processes, capricious markets, government politics and international engagements. Here, too, they have been inadequately appreciated by the sciences informing management of public health, rangelands, watersheds, soils and vegetation, which have frequently been based on ideas of predictability and managerial control.

Work in developing countries also underlines that public dissent and lack of trust in expert institutions is not so new. Local people have reflected on, responded to and resisted 'inappropriate' technologies and development plans in a variety of ways. Public experiences of science as part of the legitimation of powerful institutions dates back to early colonial times, and now thrives around forests in West Africa or water and dam development in India.[60] Nevertheless, as this work shows, the concerns around risk that animate consumers in the 'north' may be substantially different from the immediate livelihood concerns of the poor.

Risk and control

Both north and south, public reflections on science, technology and risks turn not just on the contents of science, but also on the institutions controlling it. In many recent examples, public controversies have been animated by the agendas and motivations of globalised and private sector firms. In India, farmers burning GM crops were reacting to Monsanto's control and to the vision of a private-sector-led agro-industrial future, not necessarily to the technology or the specific social and environmental impacts of GM crops per se.[61] When oral polio vaccine (OPV) was rejected as allegedly containing anti-fertility agents by several northern Nigerian states in 2003, derailing the global polio eradication campaign, concerns centred on the motivations of top-down global campaigns, amid what was perceived as national and global anti-Islamic sentiment. Spearheading the boycott were Muslim leaders who claimed that OPV contamination was part of a plot by western governments to reduce Muslim populations worldwide. These claims interplayed with international tensions around Islam and American imperialism post 9/11, as well as with long-standing tensions in Nigerian politics between the north and the south, and between federal and state government. At the same time, the high resource levels and political attention devoted to the OPV campaign contrasted so starkly with the near-collapse of Nigeria's routine immunisation and primary health care delivery that local communities suspected 'other' motivations for the campaign. Locally, anxieties about OPV drew on past incidents of alleged malpractice by the international health community. In 1996, families in Kano accused the US-based pharmaceutical company Pfizer of using an experimental meningitis drug on patients without fully informing them of the risks.[62]

Regulating technologies needs to take account of public under-standings of risk and uncertainty. It has been tempting for national and international agencies to write off public concerns that appear 'irrational', as ill-founded rumours or the expressions of an ignorant public. Yet experience has shown that treating public concerns in this way can fuel further controversy. Forums are needed which allow different views of risk and uncertainty to be deliberated on in such a way that they can inform regulatory responses.

In such processes it is critical to make the distinctions between risk (where probabilities of outcomes are known), uncertainty (where the full range of possible outcomes is not known) and ignorance (where we don't know what we don't know).[63] Science and scientists can help define risk parameters when probabilities are known, but once in the realms of uncertainty and ignorance, wider deliberation involving scientists and publics is necessary in order to map out possible futures and consequences and to consider the effects these might have on different social groups. Given that most debates about science and technology options involve uncertainty, and often ignorance, public debate about regulatory regimes is essential.

However, in contemporary practice regulations rarely emerge in this way. For developing countries access to global markets is essential for economic survival. Countries have little choice but to comply with regulations and standards set by international institutions, which in the name of 'free trade' try to standardise and harmonise globally. Uniform standards emerge for food safety from the Codex

Alimentarius commission of the UN Food and Agricultural Organization. For the export of livestock products, compliance with international standards from the World Animal Health Organization (OIE), import requirements (for example from the European Union) or the demands of large retailers is essential. Developing countries therefore have to comply with regimes set elsewhere, with limited purchase on the negotiation and standard-setting process. In some cases, developing country representatives may have a place at the table but little voice. And increasingly, international standards are being created in private settings – for example through negotiations among global supermarket chains – in which developing countries have little opportunity to play any part.

This global harmonisation of standards is justified by appeal to the idea of 'sound science', something which is assumed to apply universally. Thus many global food safety standards rely on the principle of 'substantial equivalence' whereby products are assessed in relation to their chemical composition regardless of the production process. On this basis maize, for instance, whether GM or not would be subject to the same standards.[64] Yet a different and equally scientific perspective would posit that the process of genetic modification mattered. If we bring in wider social and ecological concerns, the 'equivalence' looks increasingly spurious. Equally, standards for animal exports are currently set around the requirement of 'disease freedom', whereby particular diseases are eliminated from an area. Again, this is justified on the basis of 'sound' veterinary science and epidemiology. However, 'safe trade' may be realised in other ways, for example by the treatment of animal products prior to export without satisfying the onerous demands of disease freedom which may be impossible for pastoral production systems in large parts of Africa.[65]

Local governance

To make technology regulation work for the poor, a new perspective is required in which scientific justifications are more attuned to the conditions facing poorer producers or consumers in the developing world. This may mean abandoning the holy grail of globally harmonised standards in favour of scientifically justified, but locally meaningful, standards and regulations that ensure food safety, health control and the minimisation of biohazards. International regulation for trade then becomes less about the technical implementation of universalised science-based rules, but more a process of negotiation of what science makes sense, what risks and uncertainties are relevant, and to whom they matter. This in turn will require an opening up of global forums to such negotiation processes and increased capacities of developing countries to participate in them effectively by developing alternative versions of standards based on new scientific arguments.

At the national level a similar response is required. Developing countries are heavily reliant on external models or templates for regulation derived from European or North American settings. Whether for protocols for scientific research or the regulation of products in health or agriculture, a set of requirements is often adopted which may not fit local priorities, and may in turn not be implementable given government capacities. Thus elaborate biosafety regulations based on OECD models have become the focus for intensive 'capacity-building' exercises in developing countries, resulting in legal frameworks, monitoring and enforcement systems that often miss their mark. Clinical trials are conducted through procedures imposed by northern agencies, such as the US Federal Drug Administration, in ways that may be misinterpreted or have no purchase in the countries concerned.

This elaborate investment in regulation and regulatory capacity in developing countries makes strong assumptions about how things work. One assumption is that there is an effective governmental system backed up by enforceable law, or indeed that this can be created easily by external aid programmes. This may fundamentally misrepresent the ways states work, for instance in settings where government action is driven by complex, politicised and historically embedded patron–client relations. Attempts to forge an effective regulatory state also run counter to opposite trends, equally pushed by aid agencies, which have emphasised the downsizing of state functions, and the decentralisation and privatisation of many arenas of state activity. Quite how the handful of staff in an underresourced government office charged with food safety control over a large African country can be expected to achieve this is anyone's guess.

A second assumption is that there is a clear set of defined and registered players, operating in transparent and formal markets, to be regulated. This overlooks the chaotic manner that technologies work their way through societies and markets, in ways that are often informal and sometimes illegal. Paradoxically it is this very context that creates many of the pressing risks and uncertainties in developing countries, whether counterfeit drugs or seeds, or resistance problems to antibiotics, antiretrovirals or insect-resistant GM crops, or simply inappropriate delivery to consumers. Yet it is also this context that makes technologies so difficult to regulate, as the example in box 4 shows.

Box 4. Regulating GM crops in India?[66]

By the time the first GM crop in India was formally approved by the government in March 2002, at least 10,000 hectares (ha), across a number of states, were already planted with Bt cotton, a GM crop which is insect-resistant thanks to an inserted bacterial gene. On formal release, the government regulations stipulated an array of requirements for any new grower. The Monsanto–Mayhco joint venture company was under an obligation to monitor progress, pending a review after three years. The regulatory requirements were stringent, including the need to maintain a 'refuge' of non-GM crops so as to avoid the build-up of resistance. Quite how this was supposed to be implemented in the smallholder cotton sector in India no one knows. The mixed GM/non-GM seed packets offered on the formal market were one concession, but the company could not guarantee what happened to the seed once it was sold to numerous small-scale cotton farmers. But even four years on and multiple sublicensing agreements later, illegal Bt cotton may still be (nobody knows for sure) being planted.

This illegal Bt cotton originated in Gujarat from a Ahmedebad-based company, Navbharat Seeds. The Bt gene, almost certainly originating from Monsanto, was back-crossed into a number of varieties and sold on under different labels. Other small seed-bulking and supply companies followed suit, and soon across the cotton belt cheap, illegal Bt cotton seed was available on the market. The insect-resistant properties and low prices were attractive and many farmers began experimenting with it. That the illegal versions were marketed in varieties more suited to many farm conditions than the legal ones made the pirated products even more attractive. Officials of Monsanto and Mayhco, as well as government regulators, apparently did not know about this expansion of illegal growing. When news broke that Gujarat was covered with illegal Bt cotton, there were calls for strong action – the burning of the crop and the arrest of the perpetrators. Dr DB Desai, owner of Navbharat Seeds, was duly arrested and there was some ceremonial burning of illegal crops. But farmers, backed by the state government, resisted a wholesale destruction of the crop without very significant compensation. Despite all the newspaper commentary and government announcements, the illegal planting was a *fait accompli*. The industry representatives were surprisingly quiet. Many regarded the planting as evidence of demand for the technology, showing on the ground farmer acceptance on a large scale. This they thought bode well for their future market, and in many respects, with the area planted with approved varieties growing from 72,000 ha in 2002 to 1.3 million ha in 2004, they were right. It also meant that the regulatory stipulations of the expert committees and cross-ministerial approval body were likely to be fairly meaningless. Those planting illegally were not adhering to any such requirement, and indeed were reusing seeds with uncertain consequences for Bt expression and pest-resistance dynamics.

There remains much confusion: a formal, science-based regulatory system operates in parallel with a free-market, illegal and unregulated system. With an unregulated free-for-all, the impacts on health, environment and livelihoods remain unknown. Attempts to streamline the regulation of GM crops have been ongoing for some years, and a major report authored by Professor MS Swaminathan was released in 2005.[67] But while going some way to making the regulatory system more manageable, the reforms do not grapple with how to make regulation real on the ground.

So how can risks and uncertainties be dealt with where there is no regulation or where things appear unregulatable? Are there new ways of thinking about regulation that are more appropriate to the realities of developing country settings?

Inclusive regulation

New forms of inclusive regulation may be required, building regulatory capacities from below and through technology supply and consumption chains, in a dispersed and devolved manner that hitches onto locally relevant relations of trust. Such inclusive regulation must include ways of communicating about risk, uncertainty and ignorance in ways that make sense to the multiple actors involved, including poor people. In the context of healthcare regulation in Tanzania, a move away from conventional but ineffective rule-based regulation has been proposed, towards a concept of 'collaborative regulatory intervention'. This involves government and non-government actors – including health providers, NGO staff and people in local communities – working together to identify and build on socially valued behaviour, such as subsidising access to the poor and rejecting corruption, to value and strengthen those providers who successfully serve poor people's needs, to negotiate appropriate and legitimate rules, and to strengthen the claims of low-income patients.[68]

A range of regulatory practices has emerged in particular settings which might provide new ways forward. Branding schemes that link products with certified suppliers or trusted institutions – for example a World Health Organization seal of approval – are emerging in parallel to formal regulatory systems, as companies with bona fide products are keen to assure their market. In areas where drug markets include counterfeits, consumers have become adept at 'barcode literacy', checking the origins of products for themselves. Similar local-level regulation may emerge through the training of traders and small stockists to check the quality of the products they sell. New uses of the internet, including, for instance, the growth of internet pharmacies, may potentially present opportunities for giving consumers access to trusted technological products and information on risks and uncertainties on which they can make choices.

6. Engaging citizens

There has been a rush in recent years to increase public participation around issues involving science and technology. In some cases, this is a response to public controversies and concerns over risks from technology. In others, it is seen as a way to link science and technology better with people's needs by including citizen input to technology design and assessment. Whereas these concerns have arisen quite recently in the north, in the south they link with a long history of advocacy and action around participation in development. Today, in developed and developing country settings alike, few would deny the importance of citizen participation. But who exactly are these citizens, and what forms of engagement are appropriate?

Public participation exercises are initiated by many different types of organisation – from donor agencies and government departments to international and local NGOs. They employ different techniques – citizens' juries, consensus conferences, deliberative panels, multi-criteria mapping, and exercises using visual techniques, whether high-tech geographical information systems or low-tech sand drawings and stone piles. Box 5 illustrates some recent examples.

Box 5. Citizen participation in practice

Geographical information systems for participation in South Africa[69]
Geographical information systems (GIS) have conventionally been used as a top-down planning tool, yet recent applications illustrate

their potential for engaging citizens. In Namaqualand, South Africa, GIS have been used successfully to incorporate citizen expertise into models of water quality, and to engage local resource users in discussion of the scenarios produced from such combinations of citizen and expert data. In a workshop, community members created their own mental maps of their area's water resources and features. A spatially referenced database enabled these citizen maps to be integrated into a GIS. By overlaying the maps of different interest groups within the community, differing perceptions of the importance of resources and potential areas of conflict could be identified. These citizen maps were also overlaid and compared with those produced by a hydrological surveyor. Subsequent interviews and discussions explored the different perspectives on water quality, resources and their implications for local development indicated in the GIS. The citizen maps showed far more water points than had been identified by external agencies and to what use the water was being put, information that was largely unknown to the surveyor. The hydrologist's data on water quality were useful to the local communities, allowing them to find sources with low contamination and strengthening their case for better water supplies. Combining different datasets with the visual clarity of a map enhanced the understanding of both the local community and the surveyor, and allowed the potential for local groups to engage on a more level footing with outside agencies.

Prajateerpu – a citizen's jury around biotechnology in India[70]
Citizens' juries have been proposed as a way of allowing citizens to debate on contentious issues. The jury – selected either to represent a cross-section of society, or to bring together a marginalised, often unorganised interest group – discusses a proposal or series of options. The jury is encouraged to cross-examine a series of 'expert witnesses' who present particular positions and evidence, in order to come up with a 'verdict' – representing the people's

view or views – which can feed into wider policy deliberations. Pioneered in the North, one of the first applications in a developing country context was Prajateerpu, a citizen's jury held in Andhra Pradesh, India, in 2001. The jury, selected largely from smallholder farmers, many of whom were women, deliberated on a series of pre-prepared scenarios of future agriculture and rural development for the state. One scenario was based on the state government's 2020 Vision policy document, which advocated adoption of GM crops on a large scale. The jury rejected this option, putting forward a series of alternative ideals for the future that placed greater weight on local crop varieties and less-industrialised forms of farming and marketing, which could remain under greater local control.

Exploring farming and food futures in Zimbabwe[71]

Key lessons from Prajateerpu were incorporated into the design of a citizens' jury and scenario workshopping process in Zimbabwe in 2002 to explore rural futures, and particularly the role of biotechnologies within them. The *Izwi ne Tarisiro* ('Voice and Vision' in Shona) process was convened by a group of NGOs and parastatal organisations and established links with government and non-government actors from the start. It was broadly framed around the question: 'What do you desire to see happen in the smallholder agriculture sector in Zimbabwe by 2020?', rather than being led by particular technology options. A national scoping workshop involving 43 farmers from 16 districts, selected to represent a range of backgrounds, identified key issues. A jury of ten men and six women was selected from this group, which then – after a careful induction that demystified the jury and policy processes – interrogated 17 specialist witnesses over a week. The process led to agreement on some basic principles about the local control of food and farming, as well as the importance of indigenous knowledge, practical skills and local institutions. Moreover, participants felt they had had an unprecedented opportunity to interact directly with senior officials, and gain insights and information about the workings of the policy process that they could feed back to their communities, and act on in pressing for desired change.

At their best, such forms of citizen engagement bring together local knowledge and perspectives with more formal scientific expertise to produce 'solutions' that better fit poor people's concerns and priorities, and which respond better to uncertainties. However, as these examples begin to indicate and as the lessons of the longer history of participation in development more broadly have taught, participation is no panacea and can paradoxically disempower the poor.

First, participation in practice is a social event which involves power dynamics. These can result in poorer or marginalised groups being excluded, or being present but with no effective voice. Second, such events are often orchestrated, convened in the terms of their host institutions, whether these are local governments, aid agencies or activist NGOs. The effect is often to introduce a certain instrumentalism, where citizens are enrolled in a set of institutionally predefined agendas where 'science' or 'risk issues' are presented in a particular way. Citizens are cast as those who use or choose among a given array of options, rather than as those who might make or shape agendas derived from their own knowledge and framing of the issues.[72] Third, questions arise about the relationship between public participation exercises and wider political processes. Such exercises can, in practice, be isolated and isolating, serving more to support the status quo and divert opposing voices than to initiate broader processes of social and political transformation, or pro-poor shifts in innovation or regulatory systems. Participatory techniques can be, as Andy Stirling from Science and Technology Policy Research (SPRU at Sussex University) has argued, just as 'closed' as more conventional, non-participatory techniques of technology appraisal, such as cost–benefit analysis. 'Opening up' requires explicit attention to including the diverse knowledge and perspectives that different people – including the poor – bring to bear.[73] It requires treating participation not as a technocratic exercise, but as a fundamentally political process.

Alongside these more orchestrated forms of participation, citizens are also expressing their concerns with science and technology in other ways. Whether through the law, through the media, the internet or organised activism and protest, many forms of citizen mobilisation and movement in relation to science and technology are now evident. Taking advantage of new communications technologies, such movements often link local groups into regional and even global networks to press interests and claims, whether around the activities of governments, multinational corporations or the international community, as the examples in box 6 show.[74]

Box 6. Mobilising citizens

The Treatment Action Campaign in South Africa[75]

In a David and Goliath story, the grassroots organisation TAC (Treatment Action Campaign) in South Africa successfully fought through linked local and global networks to gain access to antiretroviral drugs for working class and poor people, taking on the global pharmaceutical industry and international patenting laws. In 2001, the country was in the midst of an HIV/AIDS epidemic but also a raging controversy within South Africa's scientific and political establishments over whether HIV was the cause of AIDS. TAC cut through this debate with a campaign centred on the perspectives and immediate concerns of poor and unemployed black women and men, many of whom were HIV positive and desperate for drugs for themselves and their children. Drawing on activist styles, symbols and songs from the earlier struggle against apartheid, TAC's mobilisation spread through schools, factories, community centres, churches, shabeens (drinking dens) and door-to-door visits in the townships. TAC also engaged with scientists, the media, the legal system, NGOs and government, using

---

sophisticated networking channels that crossed race, class, occupational and educational lines, and extended internationally in what has been dubbed 'grass-roots globalisation'. By focusing on moral imperatives TAC successfully forced drug companies to bring their prices down and it persuaded the Ministry of Health to make antiretroviral drugs more widely available. TAC's mobilisation was a struggle for poor people to gain access to life-saving drugs, but it was also a campaign to assert the rights of citizens to scientific knowledge, treatment information and the latest research findings.

Anti-dam struggles in India[76]

In India, large dams and river-linking systems, undertaken by government with international backing as large-scale technological 'solutions' to assumed problems of water scarcity, have long been a focus of mobilisation and protest. One of the longest-running anti-dam movements is the Sardar Sarovar (Narmada) movement, which has opposed the government/World Bank project to dam the Narmada river. With the leadership of NGOs and spokespeople, and through local meetings, demonstrations and campaigns on the global stage, the Narmada movement has given voice to citizens' concerns. These include the loss of forest-based livelihoods and cultural values centred on the river implied by flooding upstream of the dam; whether the dam will really help downstream issues of water uncertainty as lived and experienced by local farmers and pastoralists; and concerns about the elite, industrial and political interests that are perceived to drive large dam approaches. Linking up with similar movements across the world, the Narmada mobilisation has helped provoke a wave of questioning around the appropriateness of large-scale engineering technologies versus alternative approaches to addressing water issues that are better attuned to local ecological and social perspectives.

In other contexts, where the resources or political opportunities for organised movements are lacking, people who feel their livelihoods or wellbeing threatened by technologies express their concerns in less visible ways – perhaps through irony, satire or jokes, or through the many forms of subtle resistance, foot-dragging and sabotage that James Scott termed 'weapons of the weak'.[77] In the developing world, countless technology projects have met with opposition from local communities. Water pumps have mysteriously broken down, tree nurseries have gone up in flames and crop trials have been pulled from the ground.

These forms of mobilisation and cultural avenues of protest need to be taken more seriously as expressions of public concern. Alongside processes of 'invited' participation, these other forms suggest alternative and complementary routes to citizen engagement that could enable fuller inclusion of the range of poor people's views.

Active citizens

There is thus a need to 'move upstream'[78] to encompass broader questions about how science and technology agendas are framed, the social purposes they serve, and who stands to gain or lose from these. In these terms, 'public engagement with science' comes to be about much more than a narrow technical debate about risk or safety to dampen controversy. It is also about much more than engaging publics 'downstream', at the 'back end' of technology dissemination to promote acceptability or adapt technologies to local conditions. It could and should encompass dialogue and debate about future technology options and pathways, bringing the often expert-led approaches to horizon scanning, technology foresight and scenario planning to involve a wider range of perspectives and inputs. In this respect, the image of a stream may itself be too linear, implying that technological flows and pathways are already set, and it is just a question of where along them citizens might engage. Rather, and to continue the water metaphor, citizen engagement might be imagined as the diverse showers of rain that feed diverse possible streams of technological futures, whose outflows, like the water cycle, come round to shape further social and technological possibilities.

Such an approach to participation and citizen engagement in turn challenges mainstream ideas of 'the citizen'. Current views of science, technology and development, underpinned by the politics of liberal modernisation, tend to see citizens as passive beneficiaries of plans developed with formal scientific expertise and implemented through public sector institutions and global funds. In another version of the liberal view, gaining growing currency, citizens are seen as consumers of science and technology driven by market-led growth. Citizens are assumed to follow the market, while the liberal state provides a regulatory function which protects their safety. In contrast to both these views, we suggest that if science and technology are to be made to work for the poor then a third, more active version of citizenship is needed.

Citizens are, and need to be seen as, holders and creators of knowledge, actively engaging in the politics of science. They do this not just as individuals, but through emergent solidarities, sometimes on a global scale, that unite people more or less temporarily around particular issues, concerns and imaginations. Such engagement involves the claiming of rights and realisation of 'cognitive justice', in other words a genuine negotiation of knowledge linked to political negotiation between ways of life grounded in mutual recognition and respect.[79]

Such active citizen engagement is important everywhere. But it is even more necessary in developing country settings given evident failures of the liberal model: where states are too weak to plan effectively or tune global plans to country priorities, or where states cannot perform their assumed regulatory and protective function, liberal state–citizen 'contracts' fall apart. Added to the potentially more acute impacts of science and technology on livelihoods and health in developing country settings, positive and negative, the need for concerted citizen engagement along different lines becomes abundantly clear.

Active citizen engagement with science and technology has to be worked on, and learned. Yet this learning goes way beyond conventional approaches for fostering 'scientific citizenship'. These tend to focus on building scientific literacy – increasing public understanding of the content and processes of science – and on formal science education and communication to fill what are seen as deficits in public knowledge. Instead, stimulating active citizen–science engagement involves building the capacity of citizens for presenting and negotiating perspectives, and creating footholds for these in the political processes around science, technology and development. It involves linking issues of citizenship with processes we have discussed in earlier chapters: problem-framing, innovation, the governance of technologies and the regulation of risks and uncertainties.

If this dialogue is to work and have meaning, all parties have to buy into the process of negotiation based on mutual respect and learning. It should be remembered that scientists, administrators and policy-makers are citizens too. Cultures of science, administration and policy could be enriched if those who work within them are enabled to reflect on the social implications of their contributions, both as individuals and as part of institutions. Such reflections emerge most effectively when people are exposed to others' views and lived experience, across hierarchies, between institutions, and away from the capital city office to the urban neighbourhoods, villages and fields where poorer people create their livelihoods.

Since 2002, Makerere University, with funding from the Rockefeller Foundation, has run a field internship programme for its Masters in Public Health students, offering them the opportunity to experience life as a public servant in one of Uganda's outlying areas, and to engage first-hand with health issues as felt in poor rural communities. Initial projections were for 300 students to enrol in the programme but enthusiasm for the new learning involved was such that by March 2005, more than 2800 students had already signed up to serve as interns in local governments throughout the country. Similarly, a number of donor agencies have encouraged senior officials to spend a few days living and working with families in a poor community. A senior official who took part in the World Bank's Grass Roots Immersion programme (GRIP) in India said:

*Witnessing the life of a family that has no assurance that it can
survive until the next harvest, going to bed at 8pm because there
is no light and nothing else to do and talking with parents and
children who have no expectations that the government will
improve their lives had a remarkable effect on me.*

Such experiences can help to shift the worldview of senior policy-
makers and, at their best, help fold poor people's perspectives into
policy and practice at the highest level.[80]

People, participation and practice

Social scientists have a critical role to play in these processes of
fostering new relationships between scientists and poorer people,
both by helping to facilitate processes of learning about local cultural
settings and livelihood strategies, and by helping scientists and policy-
makers to appreciate and understand the social and political dimensions of science and technology change. It may not just be 'expert'
social scientists who are best positioned to play these brokerage roles.
'Frontline workers' – agricultural extension agents, village-based
health workers, community development agents, research field-
workers – must do this bridging between the worlds of villagers and
the worlds of scientists and bureaucrats daily. Frequently the
incentives and hierarchies that structure their work limit their ability
to put their local know-how into practice, as they are seen as being at
the bottom end of the expert-driven implementation chain. Reversing
these hierarchies, and recognising and appreciating fieldworker
creativity, can allow local insights to feed upwards to shape science
and technology priorities and policy.

In technical ministries around the world, professional scientists
often move from the lab or research station and end up dealing with
the processes of making policy. This of course means engaging with
the messy, complex, social and political issues that are part of all
policy processes. Whether this means prioritising expenditures in a
budget deal, dealing with aid donors in establishing the terms of
funded projects, making choices about negotiating strategies on trade

deals or providing advice to ministers and their advisers on health or agricultural policy, people with training in medicine, veterinary science, engineering or agronomy are regularly called on. Very often such individuals are excellent advisers, administrators and bureaucrats. But very often, particularly if pushed into such a role, they find the task overwhelming. In the world of policy-making, expertise is not just technical.

The questions that are relevant to policy are far more than just scientific: How can scientific and other understandings best be brought to bear on a policy problem? How can public consultations best be facilitated? How do political and commercial interests impinge on a policy decision? Should we build a public–private partnership? These are some of the questions that a social science understanding of policy prompts, and will help to address. A clear challenge, particularly in the developing world, is how to orient governmental policy-making towards development. This is not just a technical requirement – of sound economic policies or good governance as prescribed by the international donor world – but one that demands a more subtle set of skills. Linking scientific and technical understandings to improving field-level practice means acquiring a variety of skills and approaches seen as complementary to standard technical training. Box 7 illustrates a series of examples from Africa.

Box 7. Science in policy and practice: facilitating transitions in Africa
Veterinary science and policy processes[81]
The African Union's Institutional and Policy Support Team has been hosting a series of workshops over the last few years for veterinarians involved in policy-making from a variety of African countries. One participant commented: 'I thought all I had to do was explain the science and all would change – I was wrong.' Another reflected: 'There are so many interests around policy. It's like moving a big wheel. It's a long struggle.' The workshops have involved encouraging mid- to senior-level veterinary officials, all

with technical backgrounds, to reflect on how policies emerge and change in their own context. Through a series of mini action research projects participants made use of a simple framework for understanding policy frameworks and applied it to a variety of issues – from avian flu in ostriches in South Africa, to export markets and foot and mouth disease in Zimbabwe to managing Newcastle's disease in the small-scale poultry sector in Tanzania, to veterinary service reform and reorganisation in Ethiopia. The cases highlighted the differences between contexts, but also how, equipped with a better understanding of policy dynamics, change in favour of poorer livestock keepers might be encouraged.

From turning a blind eye to changing policy[82]

So-called frontline workers – agricultural extensionists, health practitioners and others – acquire ways of working that respond to the conditions that they face. Agricultural extension workers in Zimbabwe, for example, see how farmers regularly intercrop their fields, with complex combinations planted together. Such practices were in the past officially frowned on. But field extensionists realised that the technical guidelines were inappropriate and often turned a blind eye to what everyone agreed was a sensible practice. This stretched even to occasions when farmers would have otherwise failed their agricultural certificate, the requirements of which were set according to a strictly defined technical curriculum. Through the intermediation of a number of research groups and NGOs, and supported by key individuals in the government ministry's training section, the policy was changed, and farmer – and extension worker – practice was recognised.

Introducing the social dimensions to resource managers[83]

In relation to the management of wildlife Zimbabwe had become a leader in community-based approaches. Facilitation of such initiatives was seen to be as much about community participation, institutional development and policy as it was about wildlife ecology. As a result the University of Zimbabwe transformed its long-running MSc course in Tropical Resource Ecology to one that had both a natural science and social science component. From the late 1980s, graduates from across the region had a solid grounding in both ecology and applied social science, with teaching jointly provided between the Biology Department and the Centre for Applied Social Science, and in collaboration with a variety of government and NGO practitioners, ranging from the Department of National Parks and Wildlife Management to the Forestry Commission to the World Wide Fund for Nature.

While there are many examples of good practice, there is a danger that they remain isolated, with little impact on the gulfs that separate poor people's perspectives from expert-driven science and technology. Innovation to enhance and spread dynamics such as those highlighted in box 7 is a critical challenge. In institutional terms this challenge is not straightforward, given the ways that departments and disciplines are currently organised. New institutions are needed which bring together poor people, frontline workers, scientists, administrators and policy-makers in new ways that promote dialogue about long-term futures and technology options, about more immediate science and technology priorities, about technology adaptation to local contexts, and about risks and uncertainties and ways to regulate these. Such institutions would need to enable both open-ended and focused dialogue around particular problem areas. Some such institutions might operate at more local scales, but would need to articulate with national, regional and global equivalents, in a networked interaction.

Such institutions would have to incorporate a number of design principles, drawing lessons from the evident limitations of more technocratic institutional designs, even those with a participatory gloss. Key principles for such 'reflexive institutions' would include the need to sustain debate, allowing multiple worldviews to shape the discussions. For this to happen institutions must be seen by all as independent and trustworthy and operate in a transparent manner. This is of course difficult to achieve in many settings given the histories of lack of trust in state institutions and given the pervasive power asymmetry between local people and experts. But part of the aim of reflexivity is to acknowledge these differences and work on them. Such institutions will not be neat and tidy; they will be clumsy and complex and will evolve, through learning, over time and in unpredictable ways.

In the next chapter we therefore recommend piloting a global network of 'citizen's commissions for science and technology futures' as an essential complement to centres of excellence in science and technology.

7. The slow race to citizens' solutions

So where are the global science races heading? Are they leading to science and technology that works for the poor? By identifying the limitations, as well as the potential, of the race to the top in the global economy and the race to the universal fix, we have argued that the third, inevitably slower, race to citizens' solutions is a vital complement to these. Only this will ensure the inclusion of those left behind in the race to the top. And only this will ensure that the quest for powerful technologies is attuned to local needs and contexts. But the slow race does more than this. The slow race invites citizens' own knowledge and cultural understandings as a source of ideas and innovation. It involves citizens in the governance of science and technology. And it addresses the more mundane, yet essential, tasks of technology adaptation and delivery.

This pamphlet has asked what roles technologies play in the futures of people in developing countries and what the possibilities are for involving poorer women and men in their own technological futures. In this final chapter, we present some suggestions for how those who work in science and in development – as researchers, aid donors and policy-makers – can help this to happen.

1. How can citizens in developing countries become more involved in decisions about technology change?

Our key recommendation is the piloting of a set of 'citizens' commissions for science and technology futures'. These would not be a replacement for the centres of excellence in science and technology recommended by the Commission for Africa, but a complement to them. They would aim to generate citizen input into and reflection on what sorts of science and technology are needed, and how they should be governed.

The commissions would have a range of focuses and would address particular sectors (eg agriculture), technologies (eg nanotechnology) or policy issues (eg adapting to climate change). Some would address 'front-end' questions of agenda-setting and innovation. Others would address questions of risk, uncertainty and regulation. The commissions would vary geographically, with regional, multicountry commissions, national-level ones, and local forums all playing a role in different contexts. Standing commissions would interact with temporary ones, formed to address time-bound questions. And these commissions could make use of a variety of different media, from face-to-face 'public space' gatherings to online forums, blogs and virtual deliberative communities.[84]

To avoid becoming tokenistic and marginal, the commissions would need to connect to research, training and policy institutions at the local, national and global scale. They would also need to have authority and political weight in themselves. By naming them 'commissions' we have in mind the kind of clout carried by the British Royal Commission on Environmental Pollution, a multidisciplinary body respected for its authoritative contributions to science and environment issues. Citizens' commissions need to carry similar weight while directly involving citizens. They would need to develop deliberative procedures which attend to particular cultural traditions. Attention to issues of power, framing and representation will be critical to ensure that commission inquiries remain open to diverse citizen agendas.

The locations, topics and targets of the citizens' commissions would be defined from the bottom up to reflect local and regional needs. Possible targets might be the regulation of drugs to manage worldwide counterfeiting, nanotechnology possibilities for Africa, mobile phone access in sub-Saharan Africa or avian flu in south east Asia. But they should all be backed by governments, NGOs and donors, and woven into existing R&D contexts. The commissions would together build a practical set of criteria for the management of future research and the governance of technology.

Alongside citizens' commissions, we recommend greater citizen involvement in priority setting for science and technology more generally. Whether in research organisations such as the CGIAR system, in research funding organisations such as research councils and foundations, in national government ministries for science and technology, health or agriculture, or in programmes such as those forwarded by the UN, questions about the 'why' and 'who for' of technology investments need to be debated in ways that capture a diversity of citizen concerns. This means that alongside technology investment there needs to be investment in processes of participation, consultation and delivery. This in turn requires not just technical expertise, but social expertise in identifying and interpreting citizens' concerns and perspectives, and in facilitating processes through which citizens' agendas and experiences can engage directly.

2. What are the major research challenges ahead?

If science and technology are to work for the poor, scientific research is clearly crucial – in developing new drugs, vaccines and seeds, in identifying potential technological solutions to environmental, health and communications problems, in fine-tuning broad-based tech-nologies such as nanotechnology to developing country applications and in mitigating risks. Yet, as we have argued in this pamphlet, the science is not enough. Some of the most important research challenges ahead lie in linking technological progress with an under-standing of the conditions in which technologies will actually lead to improvements in people's lives and livelihoods.

An overarching challenge, therefore, is to foster more, and more effective, interdisciplinary, user-oriented and participatory research of various kinds. This involves creating research and innovation partnerships between scientists and potential users. It involves linking natural and technical science with social science. It involves linking the social and technical aspects of different sectors, sharing insights between agriculture, health and environmental science. It involves social scientists, from science and technology studies, development studies, political science or wherever, working together. And it involves researchers from the north and the south coming together in new partnerships.

Such interdisciplinary research needs to embrace the dynamism and complexity of real settings in the south. The speed of social, ecological and technical change is such that many conventional models which assume stable 'systems' are doomed to failure. Just as universalised technological 'solutions' often fail to fit complex realities, so too do universalised models for how technologies might be governed or linked with society.

Universities and research institutions in the north and south need to build commitments and mechanisms to promote these inter-disciplinary, participatory, locally connected agendas. This will involve new joint centres and institutional arrangements that link academic departments from different disciplines and locations with research 'users'. It will also involve exchanges – through visiting fellowships, internships and guided study periods – that enable researchers and users from different backgrounds to learn from each others' environments.

All this needs money, but much funding, whether from develop-ment donor agencies, foundations or research councils, is still divided up among outdated silos. Funding opportunities thus work against the kind of interdisciplinary interaction that is so necessary. In the UK, disciplinary divisions are mirrored by structures of academic funding through the Higher Education Funding Council for England (HEFCE), whose research assessment exercise perpetuates incentives for research excellence confined within narrow disciplinary boundaries.

There are, of course, exceptions, and the last few years have seen the take-off of some exciting and important funding initiatives which do promise support for the kind of interdisciplinary work which is needed. Cross-research council funding initiatives (such as the Tyndall Centre, and joint ESRC–MRC studentships); the ESRC–DFID joint research scheme for research into problems of poverty in developing countries; and the public engagement and bioethics programmes of the otherwise medically focused MRC and Wellcome Trust are all examples. But they run the risk of tokenism, enabling mainstream research to proceed on a business-as-usual basis on the grounds that the soft, social, public engagement 'stuff' has been dealt with. The challenge is to mainstream the social into the technical and vice versa, through genuinely interdisciplinary openness in funding.

3. How can people in developing countries respond to regulatory challenges arising from new technologies?

The building of regulatory systems in developing countries is a key challenge. There has been significant aid investment from a range of sources in promoting particular frameworks, for example around biosafety or food safety standards. But there are problems with 'regulation transfer' just as there are problems with 'technology transfer'. Such regulations are developed elsewhere for different purposes, and may not fit other settings. The global rules by which trade and regulation operate are projected as independent, rational and 'science-based'. But they are in practice contextual, political and normative. By viewing risk and safety in certain ways such approaches are laden with assumptions.

A key challenge identified by this pamphlet is for developing countries to develop workable regulations that fit their own circumstances and respond to people's priorities. This requires increasing the influence of developing country participants in standard-setting bodies. At a national level the focus needs to shift from the transfer of often inappropriate regulatory frameworks to the locally grounded development of new ones, responsive to local conditions, with some likelihood of being both accepted and enforced. Inclusive regulations will need to be built from the bottom up to respond to new risk and regulation issues.

4. What new types of professionals are needed for these challenges?

Developing countries need to build and retain scientific expertise and they need to foster top-quality science through new partnerships. The new biosciences research facility for east and central Africa at the Nairobi-based International Livestock Research Institute (ILRI) is an example of such a 'centre of excellence', and a central part of the African Union's strategy for enhancing scientific and technological capacity in Africa. But this pamphlet argues that we have to go beyond such centres. Elite science in Africa without strong and well-facilitated links beyond the lab is of little use. We argue that what is needed, perhaps even beyond new investments in science capacity, is investment in a new generation of professionals who are committed to and rewarded for cutting across the boundaries between the natural and social sciences, who can act as innovation brokers, and who can facilitate the processes by which diverse perspectives from poorer people are brought to bear on science and technology.

No matter how much good science is generated, without new professionals, technologies will not meet the needs of the poor. These 'bridging professionals' could be academics, public servants or NGO workers. Training them will require a combination of formal teaching, peer support, mentoring and exchange programmes. New cross-disciplinary postgraduate programmes in northern and southern institutions, including modules on public engagement and governance and providing experience of local realities in different countries, could help create a new cadre of professionals – scientists and social scientists – able to make the most of new investments in science and technology.

5. What are the challenges for the organisation of development aid spending?

Most aid money is funnelled through top-down organisations that are bad at learning from experience. Many in the aid business recognise the problem. And the increasing focus on budget support as a route to supporting development makes things worse. Much of the discussion in this pamphlet points to the need to integrate concerns about science and technology into the mainstream of development thinking and aid spending. This does not mean just large capital projects or flamboyant centres of excellence. The argument of this pamphlet is that science and technology can work for the poor but only if pro-poor innovation systems are supported by appropriate governance arrangements.

This requires aid organisations to be more explicit about the development pathways they are pursuing, and the potential trade-offs these involve. It also means there must be more explicit examination of the complexity of science and technology. Huge efforts have been made to ensure that environment, social and gender priorities are part of development planning processes. But science and technology have been seen as unproblematic technical inputs, so they do not receive the same attention. We have argued that we can run all three races, but each involves very different assumptions about what development is, and how to pursue it. Development agencies need to clarify and consider the means and ends of each race, and they need to ensure that they are run in ways that do not conflict with each other.

Running the slow race means, fundamentally, a commitment to pursuing development as people living in poverty themselves define it – placing their priorities and perspectives centre stage. A commitment to poor people's wellbeing and social justice will sometimes involve challenging dominant paradigms of modernisation, capitalism and globalisation. It will require policy goals in other areas – whether good governance or sustainable development, for instance – to be aligned towards these commitments. And it will require this commitment to be reflected in how organisations raise and allocate their funds, develop partnerships, and engage in lobbying and campaigning. As the global knowledge economy materialises, aid organisations need to acknowledge more explicitly what it takes to find citizens' solutions and make technology work for the poor.

Notes

1   G Brown, speaking at the Trades Union Congress conference, 13 Sept 2005.
2   See www.un.org/millenniumgoals/ (accessed 29 May 2006).
3   See www.commissionforafrica.org/index.html (accessed 30 May 2006).
4   N Dia and D Dickson, 'Africa: ministers call for strong science collaboration',
     SciDev.Net, 30 Sept 2005, available at www.scidev.net/News/index.cfm?
     fuseaction=readnews&itemid=2388&language=1 (accessed 30 May 2006).
5   House of Commons, 'The use of science in UK international development
     policy', House of Commons report, Oct 2004.
6   'Making science and technology work for the poor' has been the theme of
     several recent conferences and reviews. See, for example, A Ahmed, 'Making
     technology work for the poor: strategies and policies for African sustainable
     development', *International Journal of Technology, Policy and Management* 4, no
     1 (2004), available at: www.inderscience.com/search/index.php?action=
     record&rec_id=4563&prevQuery=&ps=10&m=or (accessed 28 May 2006).
7   C Juma and L Yee-Cheong, *Innovation: Applying knowledge in development*, UN
     Millennium Project Task Force on Science, Technology and Innovation
     (London: Earthscan, 2005).
8   C Juma (ed), *Going for Growth: Science, technology and innovation in Africa*
     (London: The Smith Institute, 2005).
9   See www.gatesfoundation.org/GlobalHealth/BreakthroughScience/
     GrandChallenges/Announcements/Announce-050627.htm (accessed 28 May
     2006).
10   J Wilsdon and R Willis, *See-through Science: Why public engagement needs to
     move upstream* (London: Demos, 2004); and J Wilsdon, B Wynne and J Stilgoe,
     The Public Value of Science: Or how to ensure that science really matters (London:
     Demos, 2005), both available at www.demos.co.uk (accessed 28 May 2006).
11   See, for example, Institute of Development Studies, 'Science and citizens: global
     and local voices', *IDS Policy Briefing* 30, May 2006, see www.ids.ac.uk/ids/
     bookshop/briefs (accessed 28 May 2006).

12   For example, different white papers on development have been underlain by quite different views.

13   MM Kaplan, 'The United Nations Conference on Science and Technology for Development', *WHO Chronicles* 33, no 12 (1979).

14   EF Schumacher, *Small Is Beautiful: Economics as if people mattered* (New York: Harper and Row, 1973).

15   Millennium Project Task Force on Science, Technology and Innovation, *Innovation: Applying knowledge in development*, UN Millennium Project, see www.unmillenniumproject.org/reports/tf_science.htm (accessed 30 May 2006).

16   T Blair in a speech to Goldman Sachs, London: 'Blair bowled over by city's biotech sharks', *Deccan Herald*, 24 Mar 2004. See also his Royal Society speech, May 2002 soon after his visit.

17   Numerous profiles have highlighted Bangalore as a 'hi-tech destination'. See, for example, 'Bangalore: Technopolis', *Business World*, 26 Feb 2001. But see also 'Booming computer sector seen as a mixed blessing', *Science*, 16 Dec 2005.

18   See TV Padma, 'Indian government says science needs rural focus', *SciDev.Net*, 4 Jan 2006, available at www.scidev.net/News/index.cfm?fuseaction= readNews&itemid=2566&language=1 (accessed 28 May 2006).

19   See Department for International Development, 'Evaluation of DFID's renewable natural resources research strategy (RNRRS) 1995–2005', June 2005, available at www.dfid.gov.uk/aboutdfid/performance/files/ev659s.pdf; see also www.dfid.gov.uk/research/srsa-consultation.pdf and www.dfid.gov.uk/research/srsa-response-final.pdf on the Strategy for Research on Sustainable Agriculture (all accessed 28 May 2006).

20   See 'DFID research funding framework, 2005–2007', available at www.dfid.gov.uk/pubs/files/researchframework/research-framework-2005.pdf (accessed 28 May 2006).

21   See www.grandchallengesgh.org/challenges.aspx?SecID=258 (accessed 28 May 2006).

22   See www.future-agricultures.org (accessed 28 May 2006); I Scoones (ed), *Dynamics and Diversity: Soil fertility management and farming livelihoods in Africa* (London: Earthscan, 2001).

23   L Mehta, *The Politics and Poetics of Water: Naturalising scarcity in western India* (Delhi: Orient Longman, 2006).

24   IDS, 'Making vaccine technologies work for the poor', *IDS Policy Briefing* 31, June 2006; JA Cassell et al, 'The social shaping of childhood vaccination practice in rural and urban Gambia: a quantitative survey of mothers based on ethnography', *Health Policy and Planning* (2006).

25   'African perspectives on science and technology for development: conversations with four African scholars', in *Building Science and Technology Capacity with African Partners: An Africa–Canada–United Kingdom exploration*, summary report (London: DFID, Office of Science and Technology, Canadian High Commission, Mar 2005).

26   See, for example, R Chambers, A Pacey and L-A Thrupp, *Farmer First: Farmer innovation and agricultural research* (London: IT Publications, 1989); and I

Scoones and J Thompson, *Beyond Farmer First* (London: IT Publications, 1994). On soil and water conservation see C Reij, I Scoones and C Toulmin (eds), *Sustaining the Soil: Indigenous soil and water conservation in Africa* (London: Earthscan, 1996).

27 B Oyelaran-Oyeyinka, 'Partnerships for building science and technology capacity in Africa: African experience', 2005, paper presented at the meeting 'Building science and technology capacity with African partners: an Africa–Canada–UK exploration', Canada House, 30 Jan–1 Feb 2005.

28 See 'African perspectives on science and technology'.

29 See, for example, B Lundval, *National Systems of Innovation: Towards a theory of innovation and interactive learning* (London: Pinter, 1992); A Hall et al, 'Why research partnerships really matter: innovation theory, institutional arrangements and implications for developing new technology for the poor', *World Development* 29, no 5 (2001).

30 A Barnett, *From Research to Poverty-reducing Innovation: A policy brief* (Brighton: Sussex Research Associates Ltd, 2004).

31 See 'African perspectives on science and technology'.

32 K Kar and K Pasteur, 'The subsidy of self-respect? Community led total sanitation. An update on recent developments', *IDS Working Papers* 257 (Brighton: Institute of Development Studies, 2005).

33 D Dickson, 'The case for a "deficit model" of science communication', *SciDev.Net*, 27 June 2005, see www.scidev.net/content/editorials/eng/the-case-for-a-deficit-model-of-science-communication.cfm (accessed 30 May 2006).

34 See cphp.uk.com (accessed 28 May 2006) for overviews of projects, and the following publications (among many others) arising from the programme: AJ Hall et al, 'From measuring impact to learning institutional lessons: an innovation systems perspective on improving the management of international agricultural research', *Agricultural Systems* 78 (2003). V Rasheed Sulaiman and AJ Hall, 'An innovation systems perspective on the restructuring of agricultural extension: evidence from India', *Outlook on Agriculture* 30, no 4 (2002); AJ Hall (ed), 'Special Edition: Innovation systems: agenda for north–south research collaboration and capacity development', *International Journal of Technology Management and Sustainable Development* 1, no 3 (2002); S Biggs and H Matsaert, 'Strengthening poverty reduction programmes using an actor oriented approach: examples from natural resources innovation systems', *Agren Network Paper* 134 (Jan 2004).

35 See www.cgiar-ilac.org/ (accessed 28 May 2006). Also recent Institutional Learning and Change (ILAC) briefs, including B Douthwaite and J Ashby, 'Innovation histories: a method for learning from experience', www.cgiar-ilac.org/downloads/Brief5Proof2.pdf (accessed 28 May 2006); M Lundy, MV Gottret and J Ashby, 'Learning alliances: an approach for building multistakeholder innovation systems', www.cgiar-ilac.org/downloads/Brief8Proof2.pdf (accessed 28 May 2006); and A Hall, L Mytelka and B Oyeyinka, 'Innovation systems: implications for agricultural policy and practice', www.cgiar-ilac.org/downloads/Brief2Proof2.pdf (accessed 28 May

2006). For an introduction see J Watts et al, 'Institutional learning and change: an introduction', ISNAR (International Service for National Agricultural Research) discussion paper no. 03-10 (2003).

36   See www.kist.ac.rw/ (accessed 29 May 2006).

37   See www.oecd.org/dataoecd/49/45/24236156.pdf (accessed 28 May 2006); 'UN body warns of growing innovation gap', SciDev.Net, 3 Oct 2005; www.uis.unesco.org/TEMPLATE/pdf/S&T/WdScienceRepTable1.pdf (accessed 28 May 2006); 'Asia "leads Europe" in science spending', SciDev.Net, 26 Jan 2006; CE Pray and D Umali-Deininger, 'The private sector in agricultural research systems: will it fill the gap?', World Development 26, no 6 (1998).

38   DFID, 'Research funding framework, 2005–7'.

39   'Support urged for US$160m plan for African science', SciDev.Net, 23 Sept 2005, www.scidev.net (accessed 28 May 2006).

40   World Bank, World Development Report (WDR) 2000/01: Attacking poverty (Washington, DC: World Bank, nd), see http://web.worldbank.org/WBSITE/EXTERNAL/TOPICS/EXTPOVERTY/0,,contentMDK:20194762~pagePK:1489 56~piPK:216618~theSitePK:336992,00.html (accessed 30 May 2006).

41   See discussion on page 10 in IDS, 'Rethinking health systems: a developmental and political economy perspective', report on a workshop, Institute of Development Studies, Brighton, UK, 21–22 Oct 2004.

42   J Chataway and J Smith, 'The International AIDS Vaccine Initiative (IAVI): is it getting new science and technology to the world's neglected majority?', World Development 34, no 1 (Jan 2006); www.iavi.org (accessed 28 May 2006).

43   J Chataway, J Smith and D Wield, 'Partnerships for building science and technology capacity in Africa: Canadian and UK experience', 2005, paper prepared for the meeting 'Building science and technology capacity with African partners'; www.ilri.cgiar.org (accessed 28 May 2006).

44   See www.aatf-africa.org/press-irlaunch.php (accessed 28 May 2006).

45   'We're running out of time', Newsweek, 30 Jan 2006; see news items at www.iavi.org/ (accessed 28 May 2006).

46   'Pro-poor public private partnerships for food and agriculture: an international dialogue', Washington, DC: International Food Policy Research Institute, 2005, see www.ifpri.org/events/conferences/2005/PPP/ppt/jeffries17.ppt#2 (accessed 28 May 2006). For a discussion of wider technology and animal health issues for Africa see I Scoones and W Wolmer, 'Livestock, disease, trade and markets: policy choices for the livestock sector in Africa', IDS Working Papers (Brighton: IDS, 2006).

47   For further information about the famous golden rice patent deal involving 70 patent holders across 32 different institutions see I Potrykus, 'Golden rice and beyond', Plant Physiology 125 (Mar 2001), available at www.plantphysiol.org/cgi/reprint/125/3/1157 (accessed 28 May 2006).

48   'A market for ideas: a survey of patents and technology', The Economist, 22 Oct 2005.

49   Quoted in The Economist, 22 Oct 2005.

50   See www.adelphicharter.org/ (accessed 28 May 2006) and
     www.iprcommission.org (accessed 28 May 2006).
51   See materials from Biological Innovation for Open Society at www.bios.net
     (accessed 28 May 2006); Bellagio meeting 'Open Source Models of
     Collaborative Innovation in the Life Sciences', Sept 2005; see www.merid.org
     (accessed 28 May 2006); and S Weber, *The Success of Open Source* (Cambridge,
     MA: Harvard University Press, 2004).
52   Z Thomas, 'Open source biotechnology', *Current Science* 88 (2005); N Steward,
     'Open source agriculture', *Information Systems for Biotechnology* (Dec 2005); S
     Herrera, 'Richard Jefferson, profile', *Nature Biotechnology* 23 (2005).
53   CK Prahalad, *The Fortune at the Bottom of the Pyramid: Eradicating poverty
     through profits* (Pennsylvania: Wharton School Publishing, 2004).
54   See 'Africa: the impact of mobile phones. Moving the debate forward', available
     at www.vodafone.com/assets/files/en/AIMP_17032005.pdf (accessed 3 June
     2006).
55   For briefings and reports, see publications from the Communications and
     Development theme at Panos, www.panos.org.uk/global/reportsection.asp?ID
     =1002 (accessed 28 May 2006). For cases on Bangladesh, see:
     www.developments.org.uk/data/issue31/loose-talk.htm (accessed 28 May
     2006); on Uganda www.panos.org.uk/global/featuredetails.asp?featureid=
     1157&ID=1002 (accessed 28 May 2006).
56   See commentary: 'Patent cases against Dr Reddy's Ranbaxy – Pfizer's victory
     not to impact Indian generica', *Business Line*, 13 Mar 2004. For company
     information see: www.ranbaxy.com (accessed 28 May 2006) and
     www.drreddys.com (accessed 28 May 2006). For a broader discussion see J
     Barton, 'TRIPS and the global pharmaceutical market', *Health Affairs* 23 (2004).
57   'Indian pharmaceuticals: good chemistry', *The Economist*, 4 Feb 2006.
58   R Ramesh, 'Cheap AIDS drugs under threat: body blow to developing states'
     fight against disease as Indian MPs ban copying of patented products',
     Guardian, 23 Mar 2005; 'Indian patent law will signal end of cheap HIV drugs',
     SciDev.Net, 23 Mar 2005; see also commentary at www.healthgap.org/
     press_releases/05/091905_HGAP_BP_India_patent_baker.html (accessed 28
     May 2006); and for commentary on the Indian pharmaceutical policy of 2002
     see www.pharmabiz.com/article/detnews.asp?articleid=11395&sectionid=46
     (accessed 28 May 2006).
59   U Beck, *Risk Society: Towards a new modernity* (London: Sage, 1992).
60   See, for example, J Fairhead and M Leach, *Science, Society and Power:
     Environmental knowledge and policy in West Africa and the Caribbean*
     (Cambridge: Cambridge University Press, 2003) and Mehta, *Politics and Poetics
     of Water*.
61   See for example, I Scoones, 'Contentious politics, contentious knowledges.
     Mobilisation against GM crops in Brazil, India and South Africa', *IDS Working
     Papers* 256 (Brighton, IDS, 2005).
62   M Yahya, 'Polio vaccines – difficult to swallow? The story of a controversy in
     Northern Nigeria', *IDS Working Papers* (Brighton: IDS, forthcoming).

63   A Stirling, 'Risk at a turning point?', *Journal of Environmental Medicine* 1 (1999).

64   E Millstone, E Brunner and S Mayer, 'Beyond "substantial equivalence"', *Nature* 401 (1999).

65   See Scoones and Wolmer, 'Livestock, disease, trade and markets'.

66   I Scoones, *Science, Agriculture and the Politics of Policy: The case of biotechnology in India* (Hyderabad: Orient Longman, 2006); E Shah, 'Local and global elites join hands: development and diffusion of Bt cotton technology in Gujarat', *Economic and Political Weekly*, 22 Oct 2005, available at www.epw.org.in/showArticles.php?root=2005&leaf=10&filename=9276&filetype=html (accessed 30 May 2006).

67   See www.ipgri.cgiar.org/publications/pubfile.asp?ID_PUB=1062 (accessed 30 May 2006).

68   See M Mackintosh and P Tibandebage, 'Inclusion by design? Rethinking health care market regulation in the Tanzanian context', *Journal of Development Studies* 39 (2002).

69   J Forrester and S Cinderby, 'Geographic information systems for participation' in M Leach, I Scoones and B Wynne (eds), *Science and Citizens: Globalization and the challenge of engagement* (London: Zed Press, 2005); S Cinderby, 'Geographic information systems (GIS) for participation: the future of environmental GIS?', *International Journal of Environment and Pollution* 11, no 3 (1999).

70   MP Pimbert and T Wakeford, *Prajateerpu: A citizens' jury/scenario workshop on food and farming futures for Andhra Pradesh, India* (London: International Institute for Environment and Development, 2002); I Scoones and J Thompson, 'Participatory processes for policy change: reflections on the Prajateerpu e-forum', *PLA Notes* 46 (2003).

71   E Rusike, 'Exploring food and farming futures in Zimbabwe: a citizens' jury and scenario workshop experiment' in Leach et al, *Science and Citizens*.

72   J Gaventa and A Cornwall, 'From users and choosers to makers and shapers: repositioning participation in social policy', *IDS Bulletin* 31, no 4 (2000).

73   A Stirling, 'Opening up or closing down? Analysis, participation and power in the social appraisal of technology' in Leach et al, *Science and Citizens*.

74   For further cases, see the Mobilization working paper series of the Citizens and Science Programme, Citizenship Development Research Centre, IDS; see www.drc-citizenship.org/ (accessed 29 May 2006).

75   S Robins, 'AIDS, science and citizenship after apartheid' in Leach et al, *Science and Citizens*.

76   Mehta, *Politics and Poetics of Water*.

77   J Scott, *Weapons of the Weak: Everyday strategies of peasant resistance* (New Haven: Yale University Press, 1985); J Scott, *Domination and the Arts of Resistance: Hidden transcripts* (New Haven: Yale University Press, 1990).

78   Wilsdon and Willis, *See-through Science*.

79   See S Visvanathan, 'Knowledge, justice and democracy' in Leach et al, *Science and Citizens*.

80   IDS, 'Immersions for policy and personal change: reflection and learning for development professionals', *IDS Policy Briefing* 22, July 2004.

81   J Keeley and I Scoones, *Understanding Environmental Policy Processes: Cases from Africa* (London: Earthscan, 2003); I Scoones and W Wolmer, *Policy Processes for Veterinary Services in Africa: A workshop report and training guide* (Nairobi: African Union Inter-African Bureau for Animal Resources (AU-IBAR), Pastoralist Communication Initiative/UN Office for the Coordination of Humanitarian Affairs (UNOCHA), 2005); W Wolmer and I Scoones, *Policy Processes in the Livestock Sector: Experiences from the African Union* (Nairobi: AU-IBAR, 2005).

82   I Scoones et al, *Hazards and Opportunities. Farming Livelihoods in Dryland Africa. Cases from Zimbabwe* (London: Zed Press, 1996).

83   See www.uz.ac.zw/science/biology/trep/MTRE_programme.html (accessed 30 May 2006).

84   A Kluth, 'Among the audience', *The Economist*, 20 Apr 2006.

DEMOS – Licence to Publish

THE WORK (AS DEFINED BELOW) IS PROVIDED UNDER THE TERMS OF THIS LICENCE ("LICENCE"). THE WORK IS PROTECTED BY COPYRIGHT AND/OR OTHER APPLICABLE LAW. ANY USE OF THE WORK OTHER THAN AS AUTHORIZED UNDER THIS LICENCE IS PROHIBITED. BY EXERCISING ANY RIGHTS TO THE WORK PROVIDED HERE, YOU ACCEPT AND AGREE TO BE BOUND BY THE TERMS OF THIS LICENCE. DEMOS GRANTS YOU THE RIGHTS CONTAINED HERE IN CONSIDERATION OF YOUR ACCEPTANCE OF SUCH TERMS AND CONDITIONS.

1. **Definitions**
   a **"Collective Work"** means a work, such as a periodical issue, anthology or encyclopedia, in which the Work in its entirety in unmodified form, along with a number of other contributions, constituting separate and independent works in themselves, are assembled into a collective whole. A work that constitutes a Collective Work will not be considered a Derivative Work (as defined below) for the purposes of this Licence.
   b **"Derivative Work"** means a work based upon the Work or upon the Work and other pre-existing works, such as a musical arrangement, dramatization, fictionalization, motion picture version, sound recording, art reproduction, abridgment, condensation, or any other form in which the Work may be recast, transformed, or adapted, except that a work that constitutes a Collective Work or a translation from English into another language will not be considered a Derivative Work for the purpose of this Licence.
   c **"Licensor"** means the individual or entity that offers the Work under the terms of this Licence.
   d **"Original Author"** means the individual or entity who created the Work.
   e **"Work"** means the copyrightable work of authorship offered under the terms of this Licence.
   f **"You"** means an individual or entity exercising rights under this Licence who has not previously violated the terms of this Licence with respect to the Work, or who has received express permission from DEMOS to exercise rights under this Licence despite a previous violation.
2. **Fair Use Rights.** Nothing in this licence is intended to reduce, limit, or restrict any rights arising from fair use, first sale or other limitations on the exclusive rights of the copyright owner under copyright law or other applicable laws.
3. **Licence Grant.** Subject to the terms and conditions of this Licence, Licensor hereby grants You a worldwide, royalty-free, non-exclusive, perpetual (for the duration of the applicable copyright) licence to exercise the rights in the Work as stated below:
   a to reproduce the Work, to incorporate the Work into one or more Collective Works, and to reproduce the Work as incorporated in the Collective Works;
   b to distribute copies or phonorecords of, display publicly, perform publicly, and perform publicly by means of a digital audio transmission the Work including as incorporated in Collective Works;
   The above rights may be exercised in all media and formats whether now known or hereafter devised. The above rights include the right to make such modifications as are technically necessary to exercise the rights in other media and formats. All rights not expressly granted by Licensor are hereby reserved.
4. **Restrictions.** The licence granted in Section 3 above is expressly made subject to and limited by the following restrictions:
   a You may distribute, publicly display, publicly perform, or publicly digitally perform the Work only under the terms of this Licence, and You must include a copy of, or the Uniform Resource Identifier for, this Licence with every copy or phonorecord of the Work You distribute, publicly display, publicly perform, or publicly digitally perform. You may not offer or impose any terms on the Work that alter or restrict the terms of this Licence or the recipients' exercise of the rights granted hereunder. You may not sublicense the Work. You must keep intact all notices that refer to this Licence and to the disclaimer of warranties. You may not distribute, publicly display, publicly perform, or publicly digitally perform the Work with any technological measures that control access or use of the Work in a manner inconsistent with the terms of this Licence Agreement. The above applies to the Work as incorporated in a Collective Work, but this does not require the Collective Work apart from the Work itself to be made subject to the terms of this Licence. If You create a Collective Work, upon notice from any Licencor You must, to the extent practicable, remove from the Collective Work any reference to such Licensor or the Original Author, as requested.
   b You may not exercise any of the rights granted to You in Section 3 above in any manner that is primarily intended for or directed toward commercial advantage or private monetary

compensation. The exchange of the Work for other copyrighted works by means of digital file-sharing or otherwise shall not be considered to be intended for or directed toward commercial advantage or private monetary compensation, provided there is no payment of any monetary compensation in connection with the exchange of copyrighted works.

c If you distribute, publicly display, publicly perform, or publicly digitally perform the Work or any Collective Works, You must keep intact all copyright notices for the Work and give the Original Author credit reasonable to the medium or means You are utilizing by conveying the name (or pseudonym if applicable) of the Original Author if supplied; the title of the Work if supplied. Such credit may be implemented in any reasonable manner; provided, however, that in the case of a Collective Work, at a minimum such credit will appear where any other comparable authorship credit appears and in a manner at least as prominent as such other comparable authorship credit.

5. **Representations, Warranties and Disclaimer**
   a By offering the Work for public release under this Licence, Licensor represents and warrants that, to the best of Licensor's knowledge after reasonable inquiry:
   i Licensor has secured all rights in the Work necessary to grant the licence rights hereunder and to permit the lawful exercise of the rights granted hereunder without You having any obligation to pay any royalties, compulsory licence fees, residuals or any other payments;
   ii The Work does not infringe the copyright, trademark, publicity rights, common law rights or any other right of any third party or constitute defamation, invasion of privacy or other tortious injury to any third party.
   b EXCEPT AS EXPRESSLY STATED IN THIS LICENCE OR OTHERWISE AGREED IN WRITING OR REQUIRED BY APPLICABLE LAW, THE WORK IS LICENCED ON AN "AS IS" BASIS, WITHOUT WARRANTIES OF ANY KIND, EITHER EXPRESS OR IMPLIED INCLUDING, WITHOUT LIMITATION, ANY WARRANTIES REGARDING THE CONTENTS OR ACCURACY OF THE WORK.

6. **Limitation on Liability.** EXCEPT TO THE EXTENT REQUIRED BY APPLICABLE LAW, AND EXCEPT FOR DAMAGES ARISING FROM LIABILITY TO A THIRD PARTY RESULTING FROM BREACH OF THE WARRANTIES IN SECTION 5, IN NO EVENT WILL LICENSOR BE LIABLE TO YOU ON ANY LEGAL THEORY FOR ANY SPECIAL, INCIDENTAL, CONSEQUENTIAL, PUNITIVE OR EXEMPLARY DAMAGES ARISING OUT OF THIS LICENCE OR THE USE OF THE WORK, EVEN IF LICENSOR HAS BEEN ADVISED OF THE POSSIBILITY OF SUCH DAMAGES.

7. **Termination**
   a This Licence and the rights granted hereunder will terminate automatically upon any breach by You of the terms of this Licence. Individuals or entities who have received Collective Works from You under this Licence, however, will not have their licences terminated provided such individuals or entities remain in full compliance with those licences. Sections 1, 2, 5, 6, 7, and 8 will survive any termination of this Licence.
   b Subject to the above terms and conditions, the licence granted here is perpetual (for the duration of the applicable copyright in the Work). Notwithstanding the above, Licensor reserves the right to release the Work under different licence terms or to stop distributing the Work at any time; provided, however that any such election will not serve to withdraw this Licence (or any other licence that has been, or is required to be, granted under the terms of this Licence), and this Licence will continue in full force and effect unless terminated as stated above.

8. **Miscellaneous**
   a Each time You distribute or publicly digitally perform the Work or a Collective Work, DEMOS offers to the recipient a licence to the Work on the same terms and conditions as the licence granted to You under this Licence.
   b If any provision of this Licence is invalid or unenforceable under applicable law, it shall not affect the validity or enforceability of the remainder of the terms of this Licence, and without further action by the parties to this agreement, such provision shall be reformed to the minimum extent necessary to make such provision valid and enforceable.
   c No term or provision of this Licence shall be deemed waived and no breach consented to unless such waiver or consent shall be in writing and signed by the party to be charged with such waiver or consent.
   d This Licence constitutes the entire agreement between the parties with respect to the Work licensed here. There are no understandings, agreements or representations with respect to the Work not specified here. Licensor shall not be bound by any additional provisions that may appear in any communication from You. This Licence may not be modified without the mutual written agreement of DEMOS and You.